INFINITE-DIMENSIONAL ANALYSIS

Operators in Hilbert Space; Stochastic Calculus
via Representations, and Duality Theory

Other World Scientific Titles by the Author

Recent Advances in Computational Sciences: Selected Papers from the International Workshop on Computational Sciences and Its Education
ISBN: 978-981-270-700-0

Non-commutative Analysis
ISBN: 978-981-3202-11-5
ISBN: 978-981-3202-12-2 (pbk)

INFINITE-DIMENSIONAL ANALYSIS

Operators in Hilbert Space; Stochastic Calculus via Representations, and Duality Theory

Palle Jorgensen

The University of Iowa, USA

James Tian

American Mathematics Society, USA

 **World Scientific**

NEW JERSEY · LONDON · SINGAPORE · BEIJING · SHANGHAI · HONG KONG · TAIPEI · CHENNAI · TOKYO

Published by

World Scientific Publishing Co. Pte. Ltd.

5 Toh Tuck Link, Singapore 596224

USA office: 27 Warren Street, Suite 401-402, Hackensack, NJ 07601

UK office: 57 Shelton Street, Covent Garden, London WC2H 9HE

Library of Congress Cataloging-in-Publication Data

Names: Jørgensen, Palle E. T., 1947–   author. | Tian, James F., author.

Title: Infinite-dimensional analysis : operators in Hilbert space ; stochastic calculus via
    representations, and duality theory / Palle Jorgensen, The University of Iowa, USA,
    James Tian, American Mathematics Society, USA.

Description: New Jersey : World Scientific, [2021] | Includes bibliographical references and index.

Identifiers: LCCN 2020041084 | ISBN 9789811225772 (hardcover) |
    ISBN 9789811225789 (ebook) | ISBN 9789811225796 (ebook other)

Subjects: LCSH: Functional analysis. | Hilbert space. | Stochastic analysis.

Classification: LCC QA320 .J637 2021 | DDC 515/.7--dc23

LC record available at https://lccn.loc.gov/2020041084

British Library Cataloguing-in-Publication Data

A catalogue record for this book is available from the British Library.

For any available supplementary material, please visit
https://www.worldscientific.com/worldscibooks/10.1142/11980#t=suppl

Desk Editor: Liu Yumeng

Typeset by Stallion Press
Email: enquiries@stallionpress.com

Printed in Singapore

Dedicated to the memory of Ola Bratteli, Richard V. Kadison, and Edward Nelson.

Preface

The present exposition combines a variety of themes from *infinite-dimensional analysis* of special relevance to applications, both in pure mathematics, as well as in related areas. The general theme we have in mind here is often referred to as infinite-dimensional calculus, or infinite-dimensional calculus of variation. Since it entails diverse topics from analysis, we have found that students will often have difficulties getting off the ground when getting started in the field; for example with a search of the current journal literature. We have even found that our current view is missing from many (most) books in the area.

We believe that there is therefore a need for such an exposition, covering and combining the relevant areas of analysis. Such a need has become especially apparent in view of recent and diverse trends in topics from representation theory, in spectral theory; and their use in such neighboring areas as quantum physics, and in the study of stochastic processes. Add to that such applications as machine learning, neural network, and stochastic optimization.

The purpose of this book is to make available to beginning graduate students, and to others, building up from the ground, some core areas of analysis which often are prerequisites for these developments. We begin with a presentation (Chapters 1 and 2) of a selection of topics from the theory of operators in Hilbert space, algebras of operators, and their corresponding spectral theory; all themes of special relevance for our purpose. We then turn to a systematic study of a selection of non-commutative themes; again of special relevance to the particular representation theory we will be needing later. For this task, we begin (in Chapter 3) with a study of representations of the canonical commutation relations (CCRs);

with an emphasis on the requirements of infinite-dimensional calculus of variations, often referred to as Itô and Malliavin calculus, Chapters 4–6.

We have included a number of features which we hope will make the book more accessible. These should also serve to make the presentation of some technical points inside the book more reader-friendly. For example, there are places where we flesh out, and articulate in more detail, the big picture. This is done with addition of reader-guides, and a few mathematical subsections with explanation of interconnections, add historical pointers, as well as pointers to diverse applications. This should help motivate better some of the technical mathematical themes that are perhaps more difficult. Overall, the book is comprised of *three main areas*, and we have aimed to make it as painless as possible for readers to transition from one to the next.

Organization of the book: A bird's eye view, and tips for the reader.

While the chapter-themes in the book follow a logical progression, some readers might want first to pick selected topics of personal interest. For this reason, we strive to make each individual chapter reasonably self-contained. Some chapters, and sections, can stand on their own, for example, Sections 1.3 *Connection to Quantum Mechanics*; 4.4 *Gaussian Hilbert space*; and 5.1 analysis with *Malliavin derivatives*.

To further help readers navigate the separate topics, we have included multiple *tables* which serve to summarize main ideas. This list includes Table 1.1: Lattices of projections in Hilbert space; Table 4.1: The positive kernel for Brownian motion, and its RKHS analysis.

We cite the literature, as needed, inside the book, and at the end of each chapter we include a brief *Guide to the literature*.

Throughout the book we have credited main results to the founders, by name, of the subject at hand. For the benefit novice readers, we have included in the *Appendix* a list of short biographical sketches.

In Chapter 1 we present the part of operator theory, covering bounded and unbounded linear transformations, stressing a systematic account of spectral theory. Because of the needs of operators in the study of representations of Lie algebras and of the Canonical Commutation Relations (CCR), the emphasis here is on unbounded operators. In our presentation, we stress connections to quantum physics, but a physics background on the part of readers is not assumed.

The material we need from representation theory is covered in Chapters 2 and 3. Chapters 4 and 5 cover stochastic analysis, and the associated infinite-dimensional calculus of Itô and Malliavin. Our approach is via *positive definite kernels*, and their associated *Reproducing Kernel Hilbert Spaces* (RKHS). In Chapters 6 and 7, we make the connection between the study of representations of the CCRs, and their use in analysis of Gaussian fields. This connection will be made with the use of an important concept from representation theory, *intertwining operators*.

The last chapter covers such applications as (i) analysis of networks of resisters, relative reproducing kernels (infinite graphs of resisters, reversible Markov processes); and (ii) use of RKHS theory for optimization problems in machine learning.

Anticipated audience: "Travel guide" for different groups of readers. Below, we call attention to the following six overlapping groups. These are the groups of potential readers the present co-authors have encountered in their own work/travels. We expect that there may be others as well.

(i) Courses which stress interdisciplinary topics; we expect a variety of course levels, including departments of mathematics, physics, and statistics.

(ii) Mathematicians familiar with functional analysis, but not with all of the diverse applications; including the multi-faceted Hilbert's Sixth Problem.

(iii) Physicists familiar with aspects of Hilbert's Sixth Problem, but who might be shaky regarding their mathematical underpinnings.

(iv) Mathematicians from anyone of many areas, and levels of sophistication, who are curious about all the "fancy talk" about quantum physics and random fields, etc;

(v) Supplementary text for students in beginning graduate courses, or topic courses (in math or in neighboring departments);

(vi) A general reader looking to broaden her/his perspective.

Acknowledgments

The co-authors thank many colleagues for helpful and enlightening discussions, especially Professors Daniel Alpay, Sergii Bezuglyi, Ilwoo Cho, Paul Muhly, Myung-Sin Song; and members, faculty and students, in the Math Physics seminar at The University of Iowa, especially Prof Wayne Polyzou.

We are extremely grateful to our editor Rochelle Kronzek for help and encouragements, and to two anonymous reviewers for helpful suggestions.

In a condensed form **Hilbert's sixth problem** (presented in the year 1900) deals with physics and probability theory. Among the 22 problems, number 6 is the least specific. Hilbert asked for a Mathematical Treatment of the Axioms of **Physics**. Axioms had worked well for the foundations of geometry. Hilbert wrote: *"To treat in the same manner, by means of axioms, those physical sciences in which already today mathematics plays an important part; in the first rank are the theory of probabilities and mechanics."*

With hindsight, the half of the problem dealing with physics has been most successful for quantum physics/quantum field theory. With further hindsight, the interconnections between the two halves of the problem have gained significance over the past century. What was not anticipated in 1900 was spectacular advances in mathematical statistics which were to follow.

Among the names on the physics side, John von Neumann and Arthur Wightman are prominent. On the probability side, Andrey Kolmogorov, Paul Pierre Lvy, and Kiyoshi Itô stand out. Major contributions, however, were made by many. See, e.g., [GK54, L52, L62, Wig85, Wig76, Wig95, Mac04, VW76].

Elaborating further, Hilbert wrote: "As to the axioms of **the theory of probabilities**, it seems to me desirable that their logical investigation should be accompanied by a rigorous and satisfactory development of the method of mean values in mathematical physics, and in particular in the kinetic theory of gases. ... Boltzmann's work on the principles of mechanics suggests the problem of developing mathematically the limiting processes, ..."

In a further explanation, Hilbert proposed two specific problems: (i) an axiomatic treatment of probability with limit theorems for the foundation of statistical physics; and (ii) the rigorous theory of limiting processes "which lead from the atomistic view to the laws of motion of continua".

> *Every kind of science, if it has only reached a certain degree of maturity, automatically becomes a part of mathematics.* "Axiomatic Thought" (1918), printed in From Kant to Hilbert, Vol. 2 by William Bragg Ewald.
> *"Fifty years ago Kurt Gödel... proved that the world of pure mathematics is inexhaustible. ... I hope that the notion of a final statement of the laws of physics will prove as illusory as the notion of a formal decision process for all mathematics. If it should turn out that the whole of physical reality can be described by a finite set of equations, I would be disappointed, I would feel that the Creator had been uncharacteristically lacking in imagination."* Freeman Dyson, in Ch. 3 : Manchester and Athens.

Contents

Abbreviations

CAR	canonical anti-commutation relations (p. 6, 180.)
CCR	canonical commutation relations (p. 28, 61, 66, 180.)
CND	conditionally negative definite (p. 100, 132.)
GNS	Gelfand–Naimark–Segal (p. 7, 66, 189.)
i.i.d.	independent identically distributed (in our case i.i.d. N(0,1) samples). (p. 10, 94, 110.)
LHS, RHS	left hand side, right hand side (p. 73, 98, 103).
ONB	orthonormal basis (p. 11, 26, 89.)
PDE	partial differential equation (p. 3, 26, 161.)
PVM	projection valued measure (p. 20, 27).
QM	quantum mechanics (p. 20, 28.)
RKHS	reproducing kernel Hilbert space (p. 51, 85.)
span	linear span, linear combinations (p. 7, 91, 168).

Chapter 1

Analysis in Hilbert Space: Linear Operators in Hilbert Space with Emphasis on the Case of Unbounded Operators

The art of doing mathematics consists in finding that special case which contains all the germs of generality. Quoted in Constance Reid, "Hilbert" (1970).
"Models should be as simple as possible, but not more so." Attributed to Einstein.

While the subject of linear transformation is vast, with many subdisciplines, and many applications, for our present purposes we have made choices as follows, each motivated by the aim of the book: spectral theory, duality rules, unbounded operators, all subjects that will play a key role in later chapters on selected topics from mathematical physics, representation theory, infinite-dimensional variational calculus, and stochastic processes. Useful supplement treatments in the literature: [BR79, GJ87, Nel58, Nel69, Sto90, Wig76].

The study of linear operators in Hilbert space is fundamental in diverse areas, both in pure mathematics, and in a host of applications. In mathematics, the applications include differential equations, partial differential operators, potential theory, geometry, calculus of variation, and dynamical systems. Applications to neighboring include quantum physics, machine learning, statistics and probability, stochastic analysis as used in for example the theory of security derivatives in financial mathematics, to mention just a few. Our present source of inspiration is primarily that of quantum mechanics (QM): In the axioms for QM, observables are modeled by self-adjoint operators in Hilbert space, generally unbounded; and the notion of spectrum from mathematics will agree with that from physics, e.g., measurement of radiation spectral lines, as well as the case of continuous spectrum.

We shall use the term linear operator synonymously with "linear transformation." Perhaps the later is used more commonly in the special case of finite dimensions, so the case of linear algebra. In this setting, one makes use of the interplay between linear transformations on the one hand, and matrices on the other. A given matrix yields a linear transformation in the obvious fashion, but not conversely. Given a linear transformation, there will be many matrix representations, one for each choice of bases in the respective vector spaces. In some cases (normal transformations), there is a choice of diagonal matrix representations, the Spectral Theorem. The set of eigenvalues is called the spectrum. While this is clear enough in finite dimensions, the corresponding notion for infinite-dimensional Hilbert space normal transformations is more subtle: Although there is a precise notion of spectrum, one must allow for cases when the spectrum is a continuum.

In infinite dimensions, there are also choices, the first one being the choice of vector space. There are many reasons for why we choose to study transformations between Hilbert spaces, or linear endomorphisms in the same Hilbert space (the case of a single operator), and they will be discussed inside the book. But a key reason is the important fact from first principles: A Hilbert space $\mathscr{H}$ is a complete normed linear space with the property that the norm-squared $\|\cdot\|^2$ is defined by a positive definite inner product. The inner product allows the notion of orthogonality. A system of orthogonal vectors which are also normalized is said to be orthonormal. With the use of Zorn's lemma one then shows that every Hilbert space $\mathscr{H}$ admits an orthonormal basis (ONB). The "basis" part of the definition yields a unique coordinate representation in the ONB for every vector in $\mathscr{H}$. Moreover, when $\mathscr{H}$ is given, any two ONBs have the same cardinality, so it is meaningful to talk about dimension. If the cardinality is countable $\aleph_0$, we say that $\mathscr{H}$ is separable.

The further advantage of the Hilbert space framework is that one then gets an extension of the finite-dimensional Spectral Theorem to the infinite-dimensional setting of Hilbert space: Every normal operator (in Hilbert space) can be diagonalized. However, to make this precise, one must be more precise with the meaning of "diagonalized." One must allow for the case when the spectrum is continuous. This entails an introduction of projection valued measure, to be discussed in Chapter 2.

Among the large number of topics from the theory of operators in Hilbert space, our treatment below is focused on those particular inter-related areas which are of direct relevance to the main focus in the book: duality themes and applications; stochastic calculus; and representation theory.

1.1 Basics of Hilbert space theory

> *"Mathematics as we know it, and as it has come to shape modern science, could never have come into being without some disregard for the dangers of the infinite."* David Bressoud (*A radical approach to real analysis*, MAA, 2007, p. 22)

Key to functional analysis is the idea of *normed vector spaces*. The interesting ones are infinite-dimensional. To use them effectively in the solution of problems, we must be able to take limits, hence the assumption of completeness. A complete normed linear space is called a *Banach space*. For applications in physics, statistics and in engineering, it often happens that the norm comes from an *inner product*; — this is the case of *Hilbert space*. With an inner product, one is typically able to get much more precise results, than in the less structured case of Banach space. (Many Banach spaces are not Hilbert spaces.)

The introduction of the axioms of Hilbert space and the resulting modern theory of linear transformations (operators) is motivated by such applications as quantum physics, calculus of variation, operator algebras, representation theory, optimization, machine learning, the theory of partial differential equations (PDEs), potential theory, and stochastic analysis (*Itô calculus and Malliavin derivatives*, see Section 3.3 and Chapter 5), to mention just a few. These, and additional applications, will be discussed in detail inside the book.

In the literature, the study of analysis in Hilbert space often goes by the name *operator theory*. The word "operator" refers to a linear transformation, say T, mapping a given Hilbert space into itself. For many applications, the following wider framework is needed. The starting point mentioned above is expanded in two ways: (i) For a given operator T, acting on a Hilbert space $\mathscr{H}$, it is important to allow for the case when T is not defined on all of $\mathscr{H}$ but only on a dense linear subspace, called the domain of T, written $dom\,(T)$. We then say that T is densely defined. (ii) For many problems, there will in fact be multiple Hilbert spaces involved, and one is then led to study linear operators acting between them. When two Hilbert spaces $\mathscr{H}_1$ and $\mathscr{H}_2$, are specified, a given operator T will then be defined in $\mathscr{H}_1$ (or a dense linear subspace of $\mathscr{H}_1$) and mapping into $\mathscr{H}_2$.

The more interesting Hilbert spaces typically arise in concrete applications as infinite dimensional spaces of function. And as such, they have proved indispensable tools in the study of partial differential equations (PDE), in quantum mechanics, in Fourier analysis, in signal processing, in representations of groups, and in ergodic theory. The term Hilbert space was originally coined by John von Neumann, who identified the axioms that now underlie these diverse applied areas. Examples include spaces of *square-integrable functions* (e.g., the L^2 random variables of a probability space), *Sobolev spaces*, Hilbert spaces of Schwartz distributions, and *Hardy spaces* of holomorphic functions; — to mention just a few.

One reason for their success is that geometric intuition from finite dimensions often carries over: e.g., the Pythagorean Theorem, the parallelogram law; and, for optimization problems, the important notion of "*orthogonal projection.*" And the idea (from linear algebra) of diagonalizing a normal matrix; — the *spectral theorem*.

Linear mappings (transformations) between Hilbert spaces are called *linear operators*, or simply "operators." They include partial differential operators (PDOs), and many others.

Definition 1.1. Let X be a vector space over $\mathbb{C}$. A norm on X is a mapping $\|\cdot\| : X \to \mathbb{C}$ such that

- $\|cx\| = |c|\,\|x\|$, $c \in \mathbb{C}$, $x \in X$;
- $\|x\| \geq 0$; $\|x\| = 0$ implies $x = 0$, for all $x \in X$;
- $\|x + y\| \leq \|x\| + \|y\|$, for all $x, y \in X$.

Let $(X, \|\cdot\|)$ be a normed space. X is called a *Banach space* if it is complete with respect to the induced metric

$$d(x, y) := \|x - y\|, \quad x, y \in X.$$

Definition 1.2 (Inner product). Let X be vector space over $\mathbb{C}$. An inner product on X is a function $\langle \cdot, \cdot \rangle : X \times X \to \mathbb{C}$ such that for all $x, y \in \mathscr{H}$, and $c \in \mathbb{C}$, we have

- $\langle x, \cdot \rangle : X \to \mathbb{C}$ is linear (linearity)
- $\langle x, y \rangle = \overline{\langle y, x \rangle}$ (conjugation)
- $\langle x, x \rangle \geq 0$; and $\langle x, x \rangle = 0$ implies $x = 0$ (positivity)

Lemma 1.3 (Cauchy–Schwarz). *Let $(X, \langle \cdot, \cdot \rangle)$ be an inner product space, then*

$$|\langle x, y \rangle|^2 \leq \langle x, x \rangle \langle y, y \rangle, \quad \forall x, y \in X. \tag{1.1}$$

Proof. By the positivity axiom in the definition of an inner product, we see that

$$\sum_{i,j=1}^{2} \overline{c_i} c_j \langle x_i, x_j \rangle = \left\langle \sum_{i=1}^{2} c_i x_i, \sum_{j=1}^{2} c_j x_j \right\rangle \geq 0, \quad \forall c_1, c_2 \in \mathbb{C};$$

i.e., the matrix

$$\begin{bmatrix} \langle x_1, x_1 \rangle & \langle x_1, x_2 \rangle \\ \langle x_2, x_1 \rangle & \langle x_2, x_2 \rangle \end{bmatrix}$$

is positive definite. Hence the above matrix has nonnegative determinant, and (1.1) follows. $\qquad\square$

Corollary 1.4. *Let $(X, \langle \cdot, \cdot \rangle)$ be an inner product space, then*

$$\|x\| := \sqrt{\langle x, x \rangle}, \quad x \in X \tag{1.2}$$

defines a norm.

Proof. It suffices to check the triangle inequality (Definition 1.2). For all $x, y \in X$, we have (with the use of Lemma 1.3):

$$\begin{aligned} \|x + y\|^2 &= \langle x + y, x + y \rangle \\ &= \|x\|^2 + \|y\|^2 + 2\Re\{\langle x, y \rangle\} \\ &\leq \|x\|^2 + \|y\|^2 + 2\|x\|\|y\| \quad \text{(by (1.1))} \\ &= (\|x\| + \|y\|)^2 \end{aligned}$$

and the corollary follows. $\qquad\square$

Definition 1.5. An inner product space $(X, \langle \cdot, \cdot \rangle)$ is called a *Hilbert space* if X is complete with respect to the metric

$$d(x, y) = \|x - y\|, \quad x, y \in X;$$

where the RHS is given by (1.2). Two vectors x, y are said to be orthogonal, denoted $x \perp y$, if $\langle x, y \rangle = 0$.

The abstract formulation of Hilbert space was invented by von Neumann in 1925. It fits precisely with the axioms of quantum mechanics (spectral lines, etc.) A few years before von Neumann's formulation, Max Born had translated Heisenberg's quantum mechanics into modern mathematics. In 1924, in a break-through paper, Heisenberg had invented quantum mechanics, but he had not been precise about the mathematics. His use of "matrices" was highly intuitive. It was only in the subsequent years, with the axiomatic language of Hilbert space, that the group of physicists and mathematicians around Hilbert in Göttingen were able to give the theory the form it now has in modern textbooks.

1.1.1 *Positive definite functions*

A host of problems in applications begin with some prescribed structure, usually in an infinite-dimensional context, an algebra, a group, a Lie algebra. We already mentioned that algebras of relevance to quantum physics are based on a choice of relations on commutators, the best known are the canonical commutation relations (CCR), and the canonical anti-commutation relations (CAR). By a representation we shall mean a homomorphism from the algebraic structure (for example the CCR algebra) into operators acting in a Hilbert space, say H. In the case of the CCRs, this will be unbounded operators, having a common dense domain in H. For additional details on *representation theory*, see Chapter 3 below.

The most versatile tool for generating representations is based on positive functionals. When normalized, these positive functionals are called *states.* There is a general mechanism which produce representations from states. It is named after *Gelfand, Naimark,* and *Segal,* and is referred to as the *GNS-construction.* (For details, see Sections 2.3, 3.1 and 7.1.) When a state is given, one gets a corresponding pre-Hilbert space, and the Hilbert space $\mathscr{H}$ for the representation will then arise as a *Hilbert completion.* With historical hindsight, one may view the GNS-construction as a special case of a more general procedure which begins with a prescribed positive definite function, see Definition 1.6 below.

The notion of positivity introduced below is often credited to N. Aronszajn [Aro50]. Function K satisfying the condition in Definition 1.6 is said to be *positive definite* (p.d.). It shall also be referred to as a *positive definite kernel.* As outlined below, to every such kernel K there is an associated Hilbert completion denoted $\mathscr{H}(K)$ and called *the reproducing kernel Hilbert space* (RKHS), characterized by having the reproducing kernel property, defined from the inner product in $\mathscr{H}(K)$: evaluation of values $f(x)$ for $f \in \mathscr{H}(K)$ are reproduced from the kernel, and the $\mathscr{H}(K)$-inner product. Its properties and applications will play an important role in subsequent chapters (see, e.g., a detailed analysis in Chapter 4 and selected applications in Chapter 8). While the ideas of the RKHS-construction are often credited to Aronszajn, there have been parallel developments in representation theory, credited to Gelfand, Naimark and Segal (GNS), and we refer to the discussion of the literature at the end of this chapter, as well as the GNS construction itself in Chapter 2 below.

Definition 1.6. Let X be an arbitrary (nonempty) set. A function $K : X \times X \to \mathbb{C}$ is said to be *positive definite*, if for all $n \in \mathbb{N}$,

$$\sum_{i,j=1}^{n} \overline{c_i} c_j K(x_i, x_j) \geq 0 \tag{1.3}$$

for all system of coefficients $c_1, \ldots, c_n \in \mathbb{C}$, and all $x_1, \ldots, x_n \in X$.

Corollary 1.7. *The Cauchy–Schwarz inequality holds for all positive definite functions.*

Proof. It follows from the proof of Lemma 1.3. $\qquad \square$

Theorem 1.8. *Let X be a set, and let $K : X \times X \to \mathbb{C}$ be a function. Then K is positive definite if and only if there is a Hilbert space $\mathscr{H}(= \mathscr{H}_K)$, and a function $\Phi : X \to \mathscr{H}$ such that*

$$K(x, y) = \langle \Phi(x), \Phi(y) \rangle_{\mathscr{H}} \tag{1.4}$$

for all $(x, y) \in X \times X$, where $\langle \cdot, \cdot \rangle_{\mathscr{H}}$ denotes the inner product in $\mathscr{H}$.
  Given a solution Φ satisfying (1.4), then we say that $\mathscr{H}$ is minimal if

$$\mathscr{H} = \overline{span}\{\Phi(x) : x \in X\}. \tag{1.5}$$

Given two minimal solutions, $\Phi_i : X \to \mathscr{H}_i$, $i = 1, 2$ (both satisfying (1.4)); then there is a unitary isomorphism $U : \mathscr{H}_1 \to \mathscr{H}_2$ such that

$$U\Phi_1(x) = \Phi_2(x) U, \tag{1.6}$$

for all $x \in X$.

When K is a given p.d. kernel, then there are typically many solutions to (1.4) in Theorem 1.8. A particular solution to (1.4) is called a *factorization* for K.

Indeed, there is a rich variety of factorizations, each one dictated by applications. We turn to this subject in Chapter 2 below, see especially formula (2.8) from Corollary 2.5 which forms the basis for our study of *transform theory* for the RKHS $\mathscr{H}(K)$.

Proof of Theorem 1.8. Given K, set

$$H_0 := \left\{ \sum_{\text{finite}} c_x \delta_x : x \in X, c_x \in \mathbb{C} \right\} = \mathrm{span}_{\mathbb{C}} \{ \delta_x : x \in X \},$$

and define a sesquilinear form on H_0 by

$$\left\langle \sum c_x \delta_x, \sum d_y \delta_y \right\rangle_K := \sum \overline{c_x} d_y K(x, y).$$

Note that, by assumption,

$$\left\| \sum c_x \delta_x \right\|_K^2 = \left\langle \sum c_x \delta_x, \sum c_x \delta_x \right\rangle_K$$
$$= \sum_{x,y} \overline{c_x} c_y K(x, y) \geq 0,$$

where all summations are finite.

However, $\langle \cdot, \cdot \rangle_K$ is in general not an inner product since the strict positivity axiom may not be satisfied. Hence one has to pass to a quotient space by letting

$$N = \{ f \in H_0 : \langle f, f \rangle_K = 0 \},$$

and set $\mathscr{H} :=$ the Hilbert completion of the quotient space H_0/N with respect to the $\|\cdot\|_K$-norm. The fact that N is really a subspace follows from the Cauchy–Schwarz inequality (1.1).

The assertion (1.6) follows immediately.   □

The Hilbert space $\mathscr{H}$ constructed this way from a given positive definite kernel K as in Definition 1.6 is called the *reproducing kernel Hilbert space* (RKHS), and denoted $\mathscr{H}(K)$. It will play an important role in Chapter 4, see especially Section 4.1 for details.

An operator U satisfying (1.6) is said to be *intertwining*; see Chapter 7 for a systematic study.

Example 1.9 (Wiener-measure). It is possible to be more explicit about choice of the pair $(\Phi, \mathscr{H})$ in Theorem 1.8, where $\varphi : X \times X \to \mathbb{C}$ is a given positive definite function.

We may in fact choose $\mathscr{H}$ to be $L^2(\Omega, \mathscr{F}, \mathbb{P})$ where $\mathbb{P}$ depends on K, and $(\Omega, \mathscr{F}, \mathbb{P})$ is a probability space. (See Theorem 4.14 and Section 4.4 for more details.) Let $X = [0, \infty)$, and consider

$$\varphi(s,t) = s \wedge t = \min(s,t), \tag{1.7}$$

see Figure 1.1. In this case, we may then take $\Omega =$ all continuous functions on $\mathbb{R}$, and $\Phi_t(\omega) := \omega(t)$, $t \in X$, $\omega \in \Omega$.

Further, the sigma-algebra $\mathscr{F}$ is generated by cylinder-sets, and $\mathbb{P}$ is the *Wiener-measure*; and $\Phi : X \to L^2(\Omega, \mathbb{P})$ is a Gaussian process with

$$\mathbb{E}_{\mathbb{P}}(\Phi(s)\,\Phi(t)) = \int_\Omega \Phi_s(\omega)\,\Phi_t(\omega)\,d\mathbb{P}(\omega) = s \wedge t.$$

The process $\{\Phi_t\}$ is called the Brownian motion; its properties include that each Φ_t is a Gaussian random variable. Figure 1.2 shows a set of sample path of the *standard Brownian motion*.

If the positive definite function φ in (1.7) is replaced with

$$\varphi_2(s,t) = s \wedge t - st = \varphi(s,t) - st, \tag{1.8}$$

we still get an associated Gaussian process $\Phi_2 : [0,1] \longrightarrow L^2(\Omega, \mathbb{P})$, but instead with

$$\mathbb{E}_{\mathbb{P}}(\Phi_2(s)\,\Phi_2(t)) = \varphi_2(s,t) = s \wedge t - st. \tag{1.9}$$

It is a *pined Brownian motion*.

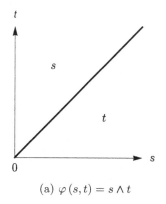

(a) $\varphi(s,t) = s \wedge t$

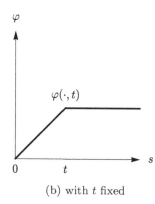

(b) with t fixed

Figure 1.1: Covariance function of Brownian motion.

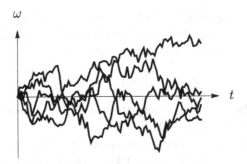

Figure 1.2: A set of Brownian sample-paths generated by a Monte-Carlo computer simulation. Each sample-path is from computer generated pseudo-random numbers, so from simulating an i.i.d. sequence of $N(0,1)$-samples.

1.1.2 *Orthonormal bases*

One of the benefits of the *Hilbert space framework* is efficiency in dealing with *orthogonal basis expansions*. However the Hilbert space framework is abstract, and one must translate into concrete cases, such as Fourier expansions, wavelet expansions, and a host of orthogonal expansions in physics, Hermite polynomials, Legendre polynomials, and more. Here we proceed to the necessary technical points. In summary, an orthonormal basis for an inner product space $\mathscr{H}$ (Hilbert space, if complete) is a basis for $\mathscr{H}$ whose vectors are orthonormal, that is, they are all unit vectors and pairwise orthogonal.

Definition 1.10. Let $\mathscr{H}$ be a Hilbert space. A family of vectors $\{u_\alpha\}$ in $\mathscr{H}$ is said to be an *orthonormal basis* if

(i) $\langle u_\alpha, u_\beta \rangle_{\mathscr{H}} = \delta_{\alpha\beta}$, and
(ii) $\overline{span}\{u_\alpha\} = \mathscr{H}$. (Here "$\overline{span}$" means "closure of the linear span.")

We are now ready to prove the existence of *orthonormal bases* for any Hilbert space. The key idea is to cook up a partially ordered set satisfying all the requirements for transfinite induction, so that each maximal element turns out to be an orthonormal basis (ONB). Notice that all we have at hand are the abstract axioms of a Hilbert space, and nothing else. Everything will be developed out of these axioms. A separate issue is *constructive* ONBs, for example, wavelets or orthogonal polynomials.

There is a new theory which generalizes the notion of ONB; called "*frame*", and it is discussed the chapters below, along with some more applications.

Theorem 1.11 (Existence of ONBs). *Every Hilbert space $\mathscr{H}$ has an orthonormal basis.*

To start out, we need the following lemma; its proof is left to the reader.

Lemma 1.12. *Let $\mathscr{H}$ be a Hilbert space and $S \subset \mathscr{H}$. Then the following are equivalent:*

(i) $x \perp S$ *implies* $x = 0$;
(ii) $\overline{span}\{S\} = \mathscr{H}$.

A set S satisfying these conditions is said to be <u>*total*</u>. *Here $x \perp S$ means $x \perp y$ for all $y \in S$.*

Corollary 1.13. *Let $\mathscr{H}$ be a Hilbert space, then $\mathscr{H}$ is isomorphic to the l^2 space of the index set of an ONB of $\mathscr{H}$. Specifically, given an ONB $\{u_\alpha\}_{\alpha \in J}$ in $\mathscr{H}$, where J is some index set, then*

$$v = \sum_{\alpha \in J} \langle u_\alpha, v \rangle \, u_\alpha, \quad and \tag{1.10}$$

$$\|v\|^2 = \sum_{\alpha \in J} |\langle u_\alpha, v \rangle|^2, \quad \forall v \in \mathscr{H}. \tag{1.11}$$

Moreover,

$$\langle u, v \rangle = \sum_{\alpha \in J} \langle u, u_\alpha \rangle \langle u_\alpha, v \rangle, \quad \forall u, v \in \mathscr{H}. \tag{1.12}$$

In Dirac's notation (see Section 1.2), (1.10)–(1.12) can be written in the following operator identity:

$$I_{\mathscr{H}} = \sum_{\alpha \in J} |u_\alpha \rangle \langle u_\alpha|. \tag{1.13}$$

(Equation (1.11) is called the Parseval identity.)

Definition 1.14. A Hilbert space $\mathscr{H}$ is said to be *separable* if it has an ONB with cardinality of $\mathbb{N}$, (this cardinal is denoted $\aleph_0$).

Many theorems stated first in the separable case also carry over to non-separable; but in the more general cases, there are both surprises, and, in some cases, substantial technical (set-theoretic) complications. As a result, we shall make the blanket assumption that our Hilbert spaces *are* separable, unless stated otherwise.

Definition 1.15. Let A be a set.

(i) Let $p : A \rightarrow \mathbb{R}_+$ be a function on A. The sum $\sum_{\alpha \in A} p(\alpha)$ is well-defined and finite, if

$$\sup_{F \subset A} \sum_{\alpha \in F} p(\alpha) < \infty$$

for all finite subsets $F \subset A$, and $\sum_{\alpha \in A} p(\alpha)$ is set to be this supremum.

(ii) By $l^2(A)$ we mean the set of functions $f : A \rightarrow \mathbb{C}$, such that

$$\sum_{\alpha \in A} |f(\alpha)|^2 < \infty.$$

Frames in Hilbert space The subject of "frames" is of relative recent vintage. It arose in signal processing, especially in the study of aliasing, and over-complete systems; see, e.g., [BLP19, Fad18, AK18, BCMS19, BCL11]. The notion of a frame of an inner product space is a generalization of a basis of a vector space to sets that may be linearly dependent. It allows one to create non-orthogonal (generalized Fourier) expansions. In signal processing, the use of frames arises in the analysis of redundant signals. In these applications, frames offer a stable way of representing a signals. Frames are used in error detection and correction and the design and analysis of filter banks, as well as in other diverse applications.

Definition 1.16. Let $\mathscr{H}$ be a Hilbert space with inner product $\langle \cdot, \cdot \rangle$. Let J be a countable index set, and let $\{w_j\}_{j \in J}$ be an indexed family of non-zero vectors in $\mathscr{H}$. We say that $\{w_j\}_{j \in J}$ is a *frame* for $\mathscr{H}$ if there are two finite positive constants b_1 and b_2 such that

$$b_1 \|u\|^2 \leq \sum_{j \in J} |\langle w_j, u \rangle|^2 \leq b_2 \|u\|^2 \tag{1.14}$$

holds for all $u \in \mathscr{H}$. We say that it is a *Parseval frame* if $b_1 = b_2 = 1$ in (1.14).

1.1.3 *Orthogonal projections in Hilbert space, and their role in probability theory*

While *orthogonal projections* in Hilbert space play a key role in a host of areas, e.g., in the theory of optimization, in quantum physics, spectral theory, and the construction of algorithms, to mention just a few, our present focus will be stochastic analysis. There orthogonal projections are conditional expectations. Indeed, then *orthogonal projections* show up as *conditional expectations*, see Chapter 4 below. First of all, any *random variable* or a system of random variables (a *stochastic process*, also called a random process) refers to an underlying probability space $(\Omega, \mathscr{C}, \mathbb{P})$. (For details, see, e.g., Sections 3.3, 4.1.2, 4.3.1, 4.4 and Theorem 4.14.)

The random variables involved represent *measurements*. Mathematically, they are measurable function with reference to an underlying *probability space*: Here Ω will be the set of *sample points* for the problem at hand, while $\mathscr{C}$ is a specification of sigma-algebra (elements in $\mathscr{C}$ are *events*). Finally, $\mathbb{P}$ will be the *probability measure* from which we are then able to define *distributions*, and *joint distributions*, referring to the particular random process. Orthogonal projections in the Hilbert space $L^2(\Omega, \mathscr{C}, \mathbb{P})$ will then be *conditional expectations*. *Reason*: The integral with respect to $\mathbb{P}$ will be an expectation, denoted $\mathbb{E}$. A choice of closed subspace in $L^2(\Omega, \mathscr{C}, \mathbb{P})$ will represent a *conditioning*. With the aid of a variety of systems of conditional expectations, one then introduces such key notions in random processes and dynamical system theory as *Markov process*, and the special processes which are *martingale*. The latter will refer to some *ordering*, for example the usual order of the time variable which might serve to index the stochastic process at hand, so conditioning future measurements with the present. Stochastic process indexed by more general sets are often called random fields.

Theorem 1.17. *Let $\mathscr{H}$ be a Hilbert space. There is a one-to-one correspondence between selfadjoint projections and closed subspaces of $\mathscr{H}$ (Figure 1.4),*

$$[\text{Closed subspace } \mathscr{M} \subset \mathscr{H}] \longleftrightarrow \text{Projections.}$$

Proof. Let P be a selfadjoint projection in $\mathscr{H}$, i.e., $P^2 = P = P^*$. Then

$$\mathscr{M} = P\mathscr{H} = \{x \in \mathscr{H} : Px = x\}$$

is a closed subspace in $\mathscr{H}$. Let $P^\perp := I - P$ be the complement of P, so that

$$P^\perp \mathscr{H} = \{x \in \mathscr{H} : P^\perp x = x\} = \{x \in \mathscr{H} : Px = 0\}.$$

Since $PP^\perp = P(1 - P) = P - P^2 = P - P = 0$, we have $P\mathscr{H} \perp P^\perp \mathscr{H}$.

Conversely, let $\mathscr{W} \subsetneq \mathscr{H}$ be a closed subspace. Note the following "parallelogram law" holds:

$$\|x + y\|^2 + \|x - y\|^2 = 2(\|x\|^2 + \|y\|^2), \quad \forall x, y \in \mathscr{H}; \tag{1.15}$$

see Figure 1.3 for an illustration.

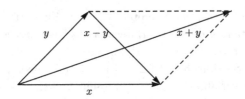

Figure 1.3: The parallelogram law.

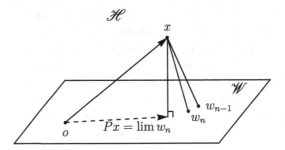

Figure 1.4: $\|x - Px\| = \inf\{\|x - w\| : w \in \mathscr{W}\}$. Projection from optimization in Hilbert space.

Let $x \in \mathscr{H} \backslash \mathscr{W}$, and set

$$d := \inf_{w \in \mathscr{W}} \|x - w\|.$$

The key step in the proof is showing that the infimum is attained; see Figure 1.4.

By definition, there exists a sequence $\{w_n\}$ in $\mathscr{W}$ so that $\|w_n - x\| \to 0$ as $n \to \infty$. Applying (1.15) to $x - w_n$ and $x - w_m$, we get

$$\|(x - w_n) + (x - w_m)\|^2 + \|(x - w_n) - (x - w_m)\|^2$$
$$= 2\left(\|x - w_n\|^2 + \|x - w_m\|^2\right);$$

which simplifies to

$$\|w_n - w_m\|^2 = 2\left(\|x - w_n\|^2 + \|x - w_m\|^2\right) - 4\left\|x - \frac{w_n + w_m}{2}\right\|^2$$

$$\leq 2\left(\|x - w_n\|^2 + \|x - w_m\|^2\right) - 4d^2. \tag{1.16}$$

Notice here all we require is $\frac{1}{2}(w_n + w_m) \in \mathscr{W}$, hence the argument carries over if we simply assume $\mathscr{W}$ is a closed convex subset in $\mathscr{H}$. We conclude from (1.16) that $\|w_n - w_m\| \to 0$, and so $\{w_n\}$ is a Cauchy sequence. Since $\mathscr{H}$ is complete, there is a unique limit,

$$Px := \lim_{n \to \infty} w_n \in \mathscr{W} \tag{1.17}$$

and

$$d = \|x - Px\| \left(= \inf_{w \in \mathscr{W}} \|x - w\| \right). \tag{1.18}$$

See Figure 1.4.

Set $P^\perp x := x - Px$. We proceed to verify that $P^\perp x \in \mathscr{W}^\perp$. By the minimizing property in (1.18), we have

$$\left\| P^\perp x \right\|^2 \le \left\| P^\perp x + tw \right\|^2$$
$$= \left\| P^\perp x \right\|^2 + |t|^2 \|w\|^2 + t \left\langle P^\perp x, w \right\rangle + \bar{t} \left\langle w, P^\perp x \right\rangle \tag{1.19}$$

for all $t \in \mathbb{C}$, and all $w \in \mathscr{W}$. Assuming $w \ne 0$ (the non-trivial case), and setting

$$t = -\frac{\left\langle w, P^\perp x \right\rangle}{\|w\|^2}$$

in (1.19), it follows that

$$0 \le -\frac{\left| \left\langle w, P^\perp x \right\rangle \right|^2}{\|w\|^2} \implies \left\langle w, P^\perp x \right\rangle = 0, \quad \forall w \in \mathscr{W}.$$

This shows that $P^\perp x \in \mathscr{W}^\perp$, for all $x \in \mathscr{H}$.

For uniqueness, suppose P_1 and P_2 both have the stated properties, then for all $x \in \mathscr{H}$, we have

$$x = P_1 x + P_1^\perp x = P_2 x + P_2^\perp x; \text{ i.e.,}$$
$$P_1 x - P_2 x = P_2^\perp x - P_1^\perp x \in \mathscr{W} \cap \mathscr{W}^\perp = \{0\}$$

thus, $P_1 x = P_2 x, \forall x \in \mathscr{H}$.

We leave the rest to the reader. See, e.g., [Rud91], [Nel69, p.62]. $\square$

Lemma 1.18 (Riesz). *There is a bijection $\mathscr{H} \ni h \longmapsto l_h$ between $\mathscr{H}$ and the space of all bounded linear functionals on $\mathscr{H}$, where*

$$l_h(x) := \langle h, x \rangle, \quad \forall x \in \mathscr{H}, \text{ and}$$
$$\|l_h\| := \sup \{|l(x)| : \|x\| \le 1\} < \infty.$$

Moreover, $\|l_h\| = \|h\|$.

Proof. Assume l is not trivial, and so $\ker(l)$ is a closed subspace of $\mathcal{H}$ with codimension 1. Choose $h \perp \ker(l)$ such that $l(h) = 1$. For all $x \in \mathcal{H}$, then

$$x = \langle h, x \rangle \, h + (x - \langle h, x \rangle \, h),$$

where the two terms on the RHS of the above equation are orthogonal, and in particular, $x - \langle h, x \rangle \in \ker(l)$. Therefore,

$$l(x - \langle h, x \rangle) = l(x) - \langle h, x \rangle \, l(h)$$
$$= l(x) - \langle h, x \rangle = 0,$$

or equivalently, $l(x) = l_h(x) = \langle h, x \rangle$. $\qquad\qquad\square$

1.1.4 *Bounded operators in Hilbert space*

> *"All stable processes we shall predict. All unstable processes we shall control."* John von Neumann.

The theory of operators in Hilbert space was developed by J. von Neumann [vN32a], M. H. Stone (among others) with view to quantum physics; and it is worth to stress the direct influence of the pioneers in quantum physics such as P. A. M. Dirac, Max Born, Werner Heisenberg, and Erwin Schrödinger. Our aim here is to stress these interconnections. While the division of the topic into the case of bounded operators vs unbounded operators in Hilbert space is convenient from the point of view of mathematics, in applications "most" operators are unbounded, i.e., they are defined and linear on suitable dense domains in the respective Hilbert spaces. There are some technical advantages of restricting the discussion to the bounded case, for example, we get normed algebras, and we avoid technical issues with domains. In both the bounded and case, and the unbounded theory, the theory is still parallel to the way we use transformations and matrices in linear algebra. But while the vector spaces of linear algebra are finite dimensional, the interesting Hilbert spaces will be *infinite dimensional*. Nonetheless we are still able to make use of matrices, only now they will be $\infty \times \infty$ matrices, and the discussion around this theme.

Definition 1.19. A bounded operator in a Hilbert space $\mathcal{H}$ is a linear mapping $T : \mathcal{H} \to \mathcal{H}$ such that

$$\|T\| := \sup \left\{ \|Tx\|_{\mathcal{H}} : \|x\|_{\mathcal{H}} \leq 1 \right\} < \infty.$$

We denote by $\mathcal{B}(\mathcal{H})$ the algebra of all bounded operators in $\mathcal{H}$.

Setting $(ST)(v) = S(T(v))$, $v \in \mathcal{H}$, $S, T \in \mathcal{B}(\mathcal{H})$, we have

$$\|ST\| \leq \|S\| \|T\|.$$

Remark 1.20. While there are multiple alternative norms in the literature, for linear operators (transformations), for our present purpose, the one from Definition 1.19 above will suffice. Of course, if some operator acts between two Hilbert spaces, its norm will depend on the choice of these Hilbert spaces. In fact, the formula for the norm in Definition 1.19 easily adapts to the case of operators transforming from one domain Hilbert spaces to a different target Hilbert space. Indeed, operators between different Hilbert spaces will play a crucial role in subsequent chapters; see, e.g., Section 1.6. But of course, *operator-norms* can only be defined for the case of bounded operators.

Corollary 1.21. *For all $T \in \mathcal{B}(\mathcal{H})$, there exists a unique operator $T^* \in \mathcal{B}(\mathcal{H})$, called the adjoint of T, such that*

$$\langle x, Ty \rangle = \langle T^*x, y \rangle, \quad \forall x, y \in \mathcal{H};$$

and $\|T^*\| = \|T\|$.

Proof. Let $T \in \mathcal{B}(\mathcal{H})$, then it follows from the Cauchy–Schwarz inequality, that

$$|\langle x, Ty \rangle| \leq \|x\| \|Ty\| \leq \|T\| \|x\| \|y\|.$$

Hence the mapping $y \longmapsto \langle x, Ty \rangle$ is a bounded linear functional on $\mathcal{H}$. By Riesz's theorem, there exists a unique $h_x \in \mathcal{H}$, such that $\langle x, Ty \rangle = \langle h_x, y \rangle$, for all $y \in \mathcal{H}$. Set $T^*x := h_x$. One checks that T^* is linear, bounded, and in fact $\|T^*\| = \|T\|$. $\square$

Definition 1.22. Let $T \in \mathcal{B}(\mathcal{H})$.

- T is *normal* if $TT^* = T^*T$;
- T is *self-adjoint* if $T = T^*$;
- T is *unitary* if $T^*T = TT^* = I_{\mathcal{H}}$ $(=$ the identity operator$)$;
- T is a (self-adjoint) *projection* if $T = T^* = T^2$.

For $T \in \mathcal{B}(\mathcal{H})$, we may write

$$R = \frac{1}{2}(T + T^*)$$

$$S = \frac{1}{2i}(T - T^*)$$

where both R and S are self-adjoint, and

$$T = R + iS.$$

This is similar to the decomposition of a complex number into its real and imaginary parts. Notice also that T is normal if and only if R and S commute. (Prove this!) Thus the study of a family of commuting normal operators is equivalent to the study of a family of commuting self-adjoint operators.

1.2   Dirac's approach

> "There is a great satisfaction in building good tools for other people to use."
>
> — Freeman Dyson

P.A.M. Dirac was very efficient with notation, and he introduced the "bra-ket" vectors [Dir35, Dir47]. This Dirac formalism has proved extraordinarily efficient, and it is widely used in physics. It deserves to be better known in the math community.

Definition 1.23. Let $\mathscr{H}$ be a Hilbert space with inner product $\langle \cdot, \cdot \rangle$. We denote by "bra" for vectors $\langle x|$ and "ket" for vectors $|y\rangle$, for $x, y \in \mathscr{H}$.

With Dirac's notation, our first observation is the following lemma.

Lemma 1.24. *Let $v \in \mathscr{H}$ be a unit vector. The operator $x \mapsto \langle v, x \rangle v$ can be written as $P_v = |v\rangle\langle v|$, i.e., a "ket-bra" vector (Definition 1.23). And P_v is a rank-one self-adjoint projection.*

Proof. First, we see that

$$P_v^2 = (|v\rangle\langle v|)(|v\rangle\langle v|) = |v\rangle\langle v, v\rangle\langle v| = |v\rangle\langle v| = P_v.$$

Also, if $x, y \in \mathscr{H}$ then

$$\langle x, P_v y \rangle = \langle x, v \rangle \langle v, y \rangle = \left\langle \overline{\langle x, v \rangle}v, y \right\rangle = \langle \langle v, x \rangle v, y \rangle = \langle P_v x, y \rangle$$

so $P_v = P_v^*$. $\qquad\square$

Corollary 1.25. *Let $F = span\{v_i\}$ with $\{v_i\}$ a finite set of orthonormal vectors in $\mathscr{H}$, then*

$$P_F := \sum_{v_i \in F} |v_i\rangle\langle v_i|$$

is the self-adjoint projection onto F.

Proof. Indeed, we have

$$P_F^2 = \sum_{v_i, v_j \in F} (|v_i\rangle\langle v_i|)(|v_j\rangle\langle v_j|) = \sum_{v_i \in F} |v_i\rangle\langle v_i| = P_F,$$

$$P_F^* = \sum_{v_i \in F} (|v_i\rangle\langle v_i|)^* = \sum_{v_i \in F} |v_i\rangle\langle v_i| = P_F,$$

and

$$P_F w = w \iff \sum_F |\langle v_i, w\rangle|^2 = \|w\|^2$$

$$\iff w \in F. \tag{1.20}$$

Since we may take the limit in (1.20), it follows that the assertion also holds if F is infinite-dimensional, i.e., $F = \overline{span}\{v_i\}$, closure. □

Remark 1.26. More generally, any rank-one operator can be written in Dirac notation as

$$|u\rangle\langle v| : \mathscr{H} \ni x \longmapsto \langle v, x\rangle u \in \mathscr{H}.$$

With the bra-ket notation, it is easy to verify that the set of rank-one operators forms an algebra, which easily follows from the fact that

$$(|v_1\rangle\langle v_2|)(|v_3\rangle\langle v_4|) = \langle v_2, v_3\rangle |v_1\rangle\langle v_4|.$$

The moment that an orthonormal basis is selected, the algebra of operators on $\mathscr{H}$ will be translated to the algebra of matrices (infinite). See Lemma 1.29.

Lemma 1.27. *Let $\{u_\alpha\}_{\alpha \in J}$ be an ONB in $\mathscr{H}$, then we may write*

$$I_{\mathscr{H}} = \sum_{\alpha \in J} |u_\alpha\rangle\langle u_\alpha|.$$

Proof. This is equivalent to the decomposition

$$v = \sum_{\alpha \in J} \langle u_\alpha, v\rangle u_\alpha, \quad \forall v \in \mathscr{H}.$$

□

A selection of ONB makes a representation of the algebra of operators acting on $\mathscr{H}$ by infinite matrices. We check that, using Dirac's notation (Definition 1.23), the algebra of operators really becomes the algebra of *infinite matrices*.

Definition 1.28. For $A \in \mathscr{B}(\mathscr{H})$, and $\{u_i\}$ an ONB, set

$$(M_A)_{i,j} = \langle u_i, Au_j\rangle_{\mathscr{H}}.$$

Most of the operators we use in mathematical physics problems are unbounded, so it is a big deal that the conclusion about matrix product is valid for unbounded operators subject to the condition that the chosen ONB is in the domain of such operators.

Lemma 1.29 (Matrix product). *Assume some ONB* $\{u_i\}_{i \in J}$ *satisfies*

$$u_i \in dom\,(A^*) \cap dom\,(B)\,;$$

then $M_{AB} = M_A M_B$, *i.e.,* $(M_{AB})_{ij} = \sum_k (M_A)_{ik}\,(M_B)_{kj}$.

Proof. By AB we mean the operator given by

$$(AB)\,(u) = A\,(B\,(u))\,.$$

Pick an ONB $\{u_i\}$ in $\mathscr{H}$, and the two operators as stated. We denote by $M_A = A_{ij} := \langle u_i, Au_j \rangle$ the matrix of A under the ONB. We compute $\langle u_i, ABu_j \rangle$.

$$
\begin{aligned}
(M_A M_B)_{ij} &= \sum_k A_{ik} B_{kj} = \sum_k \langle u_i, Au_k \rangle \langle u_k, Bu_j \rangle \\
&= \sum_k \langle A^*u_i, u_k \rangle \langle u_k, Bu_j \rangle \\
&= \langle A^*u_i, Bu_j \rangle \quad \text{[by Parseval]} \\
&= \langle u_i, ABu_j \rangle \\
&= (M_{AB})_{ij}
\end{aligned}
$$

where we used that $I = \sum |u_i \rangle \langle u_i|$. $\square$

1.3   Connection to quantum mechanics

One of the powerful applications of the theory of operators in Hilbert space, and more generally of functional analysis, is in quantum mechanics (QM) [Pol02, PK88, CP82]. Even the very formulation of the central questions in QM entails the mathematics of *unbounded self-adjoint operators*, of projection valued measures (PVM), and *unitary one-parameter groups* of operators. The latter usually abbreviated to "unitary one-parameter groups."

By contrast to what holds for the more familiar case of bounded operators, we stress that for *unbounded* self-adjoint operators, mathematical precision necessitates making a sharp distinction between the following three notions: *self-adjoint*, *essentially self-adjoint*, and *formally self-adjoint* (also called Hermitian, or symmetric). See [RS75, Nel69, vN32a, DS88]. One reason for this distinction is that *quantum mechanical observables*, momentum P, position Q, energy etc, become *self-adjoint operators* in the axiomatic language of QM. What makes it even more subtle is that these operators are both unbounded and non-commuting (take the case of P and Q which was pair from Heisenberg's pioneering paper on uncertainty.) Another subtle point entails the relationship between self-adjoint operators, projection-valued measures, and unitary one-parameter groups (as used in the dynamical description of states in QM, i.e., describing the solution of the wave equation of Schrödinger.) Unitary one-parameter groups are also used in the study of other partial differential equations, especially hyperbolic PDEs.

The discussion which follows below will make reference to this setting from QM, and it serves as motivation. The more systematic mathematical presentation of *self-adjoint operators, projection-valued measures*, and *unitary one-parameter groups* will be postponed to later in the book. We first need to develop a number of technical tools. However we have included an outline of the bigger picture in Section 1.9 (Stone's Theorem). Stone's theorem shows that the following three notions, (i) self-adjoint operator, (ii) projection-valued measure, and (iii) unitary one-parameter group, are incarnations of one and the same; i.e., when one of the three is known, anyone of the other two can be computed from it.

Much of the motivation for the axiomatic approach to the theory of linear operators in Hilbert space dates back to the early days of quantum mechanics (Planck, Heisenberg [Hei69], and Schrödinger [Sch32, Sch40, Sch99]), but in the form suggested by J. von Neumann[1] [vN31, vN32a, vN32b, vN35]. Here we will be brief, as a systematic and historical discussion is far beyond our present scope. Suffice it to mention here that what is known as the "matrix-mechanics" of Heisenberg takes the form infinite by infinite matrices with entries representing, in turn, *transition probabilities*, where "transition" refers to "jumps" between energy levels. By contrast to matrices, in Schrödinger's wave mechanics, the Hilbert space represents wave solutions to Schrödinger's equation. Now this entails the study of one-parameter groups of unitary operators in $\mathcal{H}$. In modern language, with the two settings we get the dichotomy between the case when the Hilbert space is an l^2-space (i.e., an l^2-sequence space), vs the case of Schrödinger when $\mathcal{H}$ is an L^2-space of functions on phase-space.

---

[1]von Neumann's formulation is the one now adopted by most books on functional analysis.

In both cases, the observables are represented by families of self-adjoint operators in the respective Hilbert spaces. For the purpose here, we pick the pair of self-adjoint operators representing momentum (denoted P) and position (denoted Q). In one degree of freedom, we only need a single pair. The canonical commutation relation is

$$PQ - QP = -iI; \quad \text{or } PQ - QP = -i\hbar I \qquad (1.21)$$

where $\hbar = \frac{h}{2\pi}$ is Planck's constant, and $i = \sqrt{-1}$. Here I denotes the identity operator in the Hilbert space where P and Q are realized.

A few years after the pioneering work of Heisenberg and Schrödinger, J. von Neumann and M. Stone proved that the two approaches are *unitarily equivalent*, hence they produce the same "measurements." In modern lingo, the notion of measurement takes the form of projection valued measures, which in turn are the key ingredient in the modern formulation of the spectral theorem for self-adjoint, or normal, linear operators in Hilbert space. (See [Sto90, Yos95, Nel69, RS75, DS88].) Because of dictates from physics, the "interesting" operators, such as P and Q are *unbounded*.

The first point we will discuss about the pair of operators P and Q is non-commutativity. As is typical in mathematical physics, non-commuting operators will satisfy conditions on the resulting commutators. In the case of P and Q, the commutation relation is called the canonical commutation relation; see below. For reference, see [Dir47, Hei69, vN31, vN32b].

Quantum mechanics was born during the years from 1900 to 1933. It was created to explain phenomena in black body radiation, and for the low energy levels of the hydrogen atom, where a discrete pattern occurs in the frequencies of waves in the radiation. The radiation energy turns out to be $E = \nu\hbar$, with $\hbar$ being the Plank's constant, and ν is frequency. Classical mechanics ran into trouble in explaining experiments at the atomic level.

In response to these experiments, during the years of 1925 and 1926, Heisenberg found a way to represent the energy E as a matrix (spectrum = energy levels), so that the matrix entries $\langle v_j, Ev_i \rangle$ represent transition probability for transitions from energy level i to energy level j. (See Figure 1.5 below.)

A fundamental relation in quantum mechanics is the *canonical commutation relation* satisfied by the momentum operator P and the position operator Q, see (1.22).

Definition 1.30. Given a Hilbert space $\mathscr{H}$, we say two operators P and Q in $\mathscr{B}(\mathscr{H})$ satisfy the *canonical commutation relation (CCR)*, if the following identity holds:

$$PQ - QP = -iI, \quad i = \sqrt{-1}. \tag{1.22}$$

We stress that there are no solutions to (1.22) in the form of bounded operators (a student exercise!), so implicit in the statement of (1.22) is that the two operators P and Q by necessity will be unbounded. They will have a common dense domain in the Hilbert space $\mathcal{H}$, carrying the representation. With suitable technical assumptions, it then follows that, up to unitary equivalence, there is only one irreducible representation. The same uniqueness theorem also holds for a finite number of degrees of freedom. This is the Stone–von Neumann uniqueness theorem. But for an infinite number of degrees of freedom, there is a rich variety of inequivalent irreducible representations of the CCRs, reflecting the realities of *quantum physics*. For additional details, readers are referred to Section 1.3 above, and Chapter 5 below, as well as the standard literature, for example [JT17, vN31], and the literature discussion at the end of this chapter.

Heisenberg represented the operators P, Q by infinite matrices, although his solution to (1.22) is not really matrices, and not finite matrices.

Lemma 1.31. *Equation* (1.22) *has <u>no</u> solutions for finite matrices, in fact, not even for bounded operators.*

Proof. The reason is that for matrices, there is a trace operation where

$$trace(AB) = trace(BA). \tag{1.23}$$

For (1.22), this implies the trace on the left-hand side is zero, while the trace on the RHS is not. $\qquad\square$

This shows that there is no finite dimensional solution to the commutation relation above, and one is forced to work with infinite dimensional Hilbert space and operators on it. However, Heisenberg [vN31, Hei69] found his "matrix" solutions by tri-diagonal $\infty \times \infty$ matrices, where

$$P = \frac{1}{\sqrt{2}} \begin{bmatrix} 0 & 1 & & & \\ 1 & 0 & \sqrt{2} & & \\ & \sqrt{2} & 0 & \sqrt{3} & \\ & & \sqrt{3} & 0 & \ddots \\ & & & \ddots & \ddots \end{bmatrix} \tag{1.24}$$

and

$$Q = \frac{1}{i\sqrt{2}} \begin{bmatrix} 0 & 1 & & & \\ -1 & 0 & \sqrt{2} & & \\ & -\sqrt{2} & 0 & \sqrt{3} & \\ & & -\sqrt{3} & 0 & \ddots \\ & & & \ddots & \ddots \end{bmatrix}. \tag{1.25}$$

The complex i in front of Q is to make it self-adjoint.

Remark 1.32 (Matrices vs operators). Every bounded linear operator (and many unbounded operators too) in separable Hilbert space (and, in particular, in l^2) can be realized as a well-defined *infinite "square" matrix*. In l^2 we pick the canonical ONB, but in a general Hilbert space, a choice of ONB must be made.

We saw that most rules for finite matrices carry over to the case of infinite matrices; sums, products, and adjoints. For instance, in order to find the matrix of the sum of two bounded operators, just find the sum of the matrices of these operators. And the matrix of the adjoint operator A^* (of a bounded operator A in Hilbert space) is the adjoint matrix (conjugate transpose) of the matrix of the operator A. While it is "easy" to go from bounded operators to infinite "square" matrices, the converse is much more subtle.

The representation of the canonical pair of operators P and Q as infinite by infinite matrices, see (1.24) and (1.25), is the reason for what is referred to as *Heisenberg's matrix mechanics*. The matrix view point contrasts the approach of Schrödinger with the *Schrödinger equation* (a PDE), and so Hilbert spaces of functions, representing the quantum states. For more on quantum states, measurements, and probability distributions, see Section 1.4 below.

Notice that P, Q do not commute, and the above commutation relation leads to the uncertainty principle (Hilbert, Max Born, von Neumann worked out the mathematics). It states that the statistical variance $\triangle P$ and $\triangle Q$ satisfy

$$\triangle P \triangle Q \geq \frac{\hbar}{2}. \tag{1.26}$$

We will show that non-commutativity always yields "uncertainty."

Theorem 1.33 (Uncertainty Principle). *Let $\mathscr{D}$ be a dense subspace in $\mathscr{H}$, and A, B be two Hermitian operators such that $A, B : \mathscr{D} \hookrightarrow \mathscr{D}$ (i.e., $\mathscr{D}$*

is assumed invariant under both A and B.) Then,

$$\|Ax\| \, \|Bx\| \geq \frac{1}{2} \, |\langle x, [A, B] \, x \rangle| \,, \quad \forall x \in \mathscr{D}; \tag{1.27}$$

where $[A, B] := AB - BA$ is the commutator of A and B.
   In particular, setting

$$A_1 := A - \langle x, Ax \rangle$$

$$B_1 := B - \langle x, Bx \rangle$$

then A_1, B_1 are Hermitian, and

$$[A_1, B_1] = [A, B] \,.$$

Therefore,

$$\|A_1 x\| \, \|B_1 x\| \geq \frac{1}{2} \, |\langle x, [A, B] \, x \rangle| \,, \quad \forall x \in \mathscr{D}.$$

Suppose A and B represent quantum mechanical observables. We say that they form a dual pair if their commutator is a scalar, say s. Taking squares on both of the last inequality; we get a precise statement about uncertainty as follows: The product of uncertainty in measurements of the dual observables, performed in a fixed state (here a fixed normalized vector x), will have a lower bound $|s| / 2$. Hence, the lower bound is independent of the state. (Of course, in the classical version of Heisenberg, the scalar s will be a multiple of Planck's constant. See Corollary 1.34 below.)

Proof of Theorem 1.33. By the Cauchy–Schwarz inequality (Lemma 1.3), and for $x \in \mathscr{D}$, we have

$$\|Ax\| \, \|Bx\| \geq |\langle Ax, Bx \rangle|$$

$$\geq |\Im \{\langle Ax, Bx \rangle\}|$$

$$= \frac{1}{2} \left| \langle Ax, Bx \rangle - \overline{\langle Ax, Bx \rangle} \right|$$

$$= \frac{1}{2} |\langle Ax, Bx \rangle - \langle Bx, Ax \rangle|$$

$$= \frac{1}{2} |\langle x, [AB - BA] \, x \rangle| \,. \qquad \square$$

Corollary 1.34. *If $[A, B] = i\hbar I$, $\hbar \in \mathbb{R}_+$, and $\|x\| = 1$; then*

$$\sigma_x(A) \, \sigma_x(B) \geq \frac{\hbar}{2}, \tag{1.28}$$

where the variance of T is defined as

$$\sigma_x(T) := \sqrt{\|Tx\|^2 - |\langle x, Tx \rangle|^2}, \quad x \in dom(T).$$

1.4 Probabilistic interpretation of Parseval's formula for Hilbert space

"If you are receptive and humble, mathematics will lead you by the hand." Paul Dirac As quoted in The Strangest Man: The Hidden Life of Paul Dirac, Mystic of the Atom (2009) by Graham Farmelo, p. 435.

The considerations below will serve as justification of Hilbert space for problems in quantum physics, and in probability. In particular they justify the following dictionary of "translations": Quantum states → norm-one vectors in Hilbert space. Quantum observables → self-adjoint operators. Absolute-square of inner products → transition probability. Measurements → projection-valued measures. Quantum logic gates → lattices of projections, and systems of unitary operators.

Case 1. Let $\mathscr{H}$ be a complex Hilbert space, and let $\{u_k\}_{k \in \mathbb{N}}$ be an ONB, then Parseval's formula reads:

$$\langle v, w \rangle_{\mathscr{H}} = \sum_{k \in \mathbb{N}} \langle v, u_k \rangle \langle u_k, w \rangle, \quad \forall v, w \in \mathscr{H}. \tag{1.29}$$

Translating this into a statement about "transition probabilities" for quantum states, $v, w \in \mathscr{H}$, with $\|v\|_{\mathscr{H}} = \|w\|_{\mathscr{H}} = 1$, we get

$$\text{Prob}(v \to w) = \sum_{k \in \mathbb{N}} \text{Prob}(v \to u_k) \text{Prob}(u_k \to w). \tag{1.30}$$

See Figure 1.5. The states v and w are said to be *uncorrelated* if they are orthogonal.

Fix a state $w \in \mathscr{H}$, then

$$\|w\|^2 = \sum_{k \in \mathbb{N}} |\langle u_k, w \rangle|^2 = 1.$$

The numbers $|\langle u_i, w \rangle|^2$ represent a probability distribution over the index set, where $|\langle u_k, w \rangle|^2$ is the probability that the quantum system is in the state $|u_k\rangle$.

Disclaimer: The notation "transition probability" in (1.30) and Figure 1.5 is a stretch since the inner products $\langle v, u_k \rangle$ are not positive. Nonetheless,

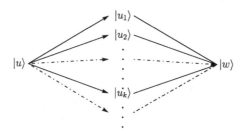

Figure 1.5: Transition of quantum-states.

it is justified by

$$\sum_k |\langle v, u_k \rangle|^2 = 1$$

when $v \in \mathscr{H}$ (is a state vector).

Case 2. If $P : \mathcal{B}(\mathbb{R}) \to \text{Proj}(\mathscr{H})$ is a projection valued measure (Definition 1.35), we get the analogous assertions, but with integration, as opposed to summation. In this case (1.29) holds in the following form:

$$\langle v, w \rangle_{\mathscr{H}} = \int_{\mathbb{R}} \langle v, P(d\lambda) w \rangle_{\mathscr{H}}; \tag{1.31}$$

and for $v = w$, it reads:

$$\|v\|^2_{\mathscr{H}} = \int_{\mathbb{R}} \|P(d\lambda) v\|^2_{\mathscr{H}}. \tag{1.32}$$

Definition 1.35 (Projection valued measure). Let $\mathcal{B}(\mathbb{R})$ be the Borel σ-algebra of subsets of $\mathbb{R}$. Let $\mathscr{H}$ be a Hilbert space.

We say $P : \mathcal{B}(\mathbb{R}) \to \text{Proj}(\mathscr{H})$ is a projection valued measure (PVM), if the following conditions are satisfied:

(i) $P(A) = P(A)^* = P(A)^2$, $\forall A \in \mathcal{B}(\mathbb{R})$.
(ii) $P(\cdot)$ is countably additive on the Borel subsets of $\mathbb{R}$, i.e.,

$$\sum_j P(A_j) = P(\cup_j A_j)$$

where $A_j \in \mathcal{B}(\mathbb{R})$, $A_i \cap A_j = \emptyset$, $i \neq j$.
(iii) $P(A \cap B) = P(A) P(B)$, $\forall A, B \in \mathcal{B}(\mathbb{R})$.

It is customary, when only axioms (i) and (ii) hold, to refer to P as a PVM. If a particular projection valued measure P should also satisfy condition (iii), then P is said to be *orthogonal*. In fact, every PVM in a fixed Hilbert space has an orthogonal realization in a "bigger" Hilbert space; we say that P has a (so called) orthogonal dilation.

1.5   The lattice structure of projections

A lattice is a partially ordered set in which every two elements have a supremum (also called a least upper bound or join) and an infimum (also called a greatest lower bound or meet).

The purpose of the discussion below is twofold; one to identify two cases: (i) the (easy) *lattice of subsets* of a fixed total set; and (ii) the *lattice of projections* in a fixed Hilbert space. Secondly we point out how non-commutativity of projections makes the comparison of (i) and (ii) subtle; even though there are some intriguing correspondences; see Table 1.1 for illustration.

In this section we will denote projections P, Q, etc.

Remark 1.36. Caution about use of the symbols P and Q. Traditionally, in the study of Hilbert space analysis, projections are denoted by the letters P and Q. This is the case also in the present section. For pairs of projections, or for systems of projections, the goal is to make comparisons precise, and to prove structure theorems. This analysis is often motivated by quantum theory.

Ironically, in Section 1.3 the *canonical pairs* of operators in quantum mechanics are denoted P and Q, P for *momentum*, and Q for *position*. Now in this context P and Q are not projections (not even bounded), but rather unbounded symmetric operators, assumed to satisfy the CCRs. Choices of self-adjoint extensions (see, e.g., Theorem 1.69) will then be an important question to consider. By the conventions from QM, momentum operators are denoted P and their dual, position operators, are denoted Q. We will have a precise formulation of *Heisenberg's uncertainty principle*. Hence we shall be interested in self-adjoint realizations (representations) for the CCRs as QM *observables* are realized by self-adjoint operators. *Measurements* entail projection valued measures. Section 1.3 above deals with QM and one degree of freedom, while our main theme in this book, see Chapters 2 and 3, will deal with an infinite number of degrees of freedom, so an infinitely indexed pair of QM canonical operators P and Q, subject to the CCRs.

von Neumann invented the notion of abstract Hilbert space in 1928 as shown in one of the earliest papers.[2] His work was greatly motivated by quantum mechanics. In order to express quantum mechanics logic

---

[2] Earlier authors, Schmidt and Hilbert, worked with infinite bases, and $\infty \times \infty$ matrices.

Table 1.1: Lattice of projections in Hilbert space, Abelian vs non-Abelian.

SETS	CHAR	PROJ	DEF
$A \cap B$	$\chi_A \chi_B$	$P \wedge Q$	$P\mathscr{H} \cap Q\mathscr{H}$
$A \cup B$	$\chi_{A \cup B}$	$P \vee Q$	$\overline{span}\{P\mathscr{H} \cup Q\mathscr{H}\}$
$A \subset B$	$\chi_A \chi_B = \chi_A$	$P \leq Q$	$P\mathscr{H} \subset Q\mathscr{H}$
$A_1 \subset A_2 \subset \cdots$	$\chi_{A_i}\chi_{A_{i+1}} = \chi_{A_i}$	$P_1 \leq P_2 \leq \cdots$	$P_i\mathscr{H} \subset P_{i+1}\mathscr{H}$
$\bigcup_{k=1}^{\infty} A_k$	$\chi_{\cup_k A}$	$\vee_{k=1}^{\infty} P_k$	$\overline{span}\{\bigcup_{k=1}^{\infty} P_k\mathscr{H}\}$
$\bigcap_{k=1}^{\infty} A_k$	$\chi_{\cap_k A_k}$	$\wedge_{k=1}^{\infty} P_k$	$\bigcap_{k=1}^{\infty} P_k\mathscr{H}$
$A \times B$	$(\chi_{A \times X})(\chi_{X \times B})$	$P \otimes Q$	$P \otimes Q \in proj(\mathscr{H} \otimes \mathscr{K})$

operations, he created lattices of projections, so that everything we do in set theory with set operation has a counterpart in the operations of projections (Table 1.1).

For example, if P and Q are two projections in $\mathscr{B}(\mathscr{H})$, then

$$P\mathscr{H} \subset Q\mathscr{H} \tag{1.33}$$

$$\Updownarrow$$

$$P = PQ \tag{1.34}$$

$$\Updownarrow$$

$$P \leq Q. \tag{1.35}$$

This is similar to the following equivalence relation in set theory

$$A \subset B \tag{1.36}$$

$$\Updownarrow$$

$$A = A \cap B. \tag{1.37}$$

In general, product and sum of projections are not projections. But if $P\mathscr{H} \subset Q\mathscr{H}$ then the product PQ is in fact a projection. Taking adjoint in (1.34) yields

$$P^* = (PQ)^* = Q^*P^* = QP.$$

It follows that $PQ = QP = P$, i.e., *containment of subspaces implies the corresponding projections commute.*

Two decades before von Neumann developed his Hilbert space theory, Lebesgue developed his integration theory [Leb05] which extends the classical Riemann integral. The monotone sequence of sets $A_1 \subset A_2 \subset \cdots$

in Lebesgue's integration theory also has a counterpart in the theory of Hilbert space.

Lemma 1.37. *Let P_1 and P_2 be orthogonal projections acting on $\mathscr{H}$, then*

$$P_1 \le P_2 \iff \|P_1 x\| \le \|P_2 x\|, \quad \forall x \in \mathscr{H} \tag{1.38}$$

$$\Updownarrow$$

$$P_1 = P_1 P_2 = P_2 P_1 \tag{1.39}$$

(*see Table 1.1.*)

Proof. Indeed, for all $x \in \mathscr{H}$, we have

$$\|P_1 x\|^2 = \langle P_1 x, P_1 x \rangle = \langle x, P_1 x \rangle = \langle x, P_2 P_1 x \rangle \le \|P_1 P_2 x\|^2 \le \|P_2 x\|^2. \quad \square$$

Theorem 1.38. *For every monotonically increasing sequence of projections $P_1 \le P_2 \le \cdots$, let*

$$P := \vee P_k = \lim_k P_k,$$

then P defines a projection, the limit.

Proof. The assumption $P_1 \le P_2 \le \cdots$ implies that $\{\|P_k x\|\}_{k=1}^{\infty}$, $x \in \mathscr{H}$, is a monotone increasing sequence in $\mathbb{R}$, and the sequence is bounded by $\|x\|$, since $\|P_k x\| \le \|x\|$, for all $k \in \mathbb{N}$. Therefore the sequence $\{P_k\}_{k=1}^{\infty}$ converges to $P \in \mathscr{B}(\mathscr{H})$ (strongly), and P really defines a self-adjoint projection. (We use "$\le$" to denote the lattice operation on projection.) Note the convergence refers to the strong operator topology, i.e., for all $x \in \mathscr{H}$, there exists a vector, which we denote by Px, so that $\lim_k \|P_k x - Px\| = 0$. $\square$

The usual Gram–Schmidt process can now be formulated in the lattice of projections.

Definition 1.39. For a linearly independent subset $\{u_k\} \subset \mathscr{H}$, the Gram–Schmidt process yields an orthonormal set $\{v_k\} \subset \mathscr{H}$, with $v_1 := u_1/\|u_1\|$, and

$$v_{n+1} := \frac{u_{n+1} - P_n u_{n+1}}{\|u_{n+1} - P_n u_{n+1}\|}, \quad n = 1, 2, \ldots;$$

where P_n is the orthogonal projection on the n-dimensional subspace

$$V_n := span\{v_1, \ldots, v_n\}.$$

See Figure 1.6.

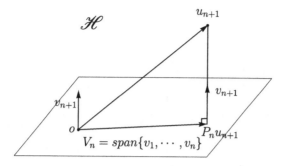

Figure 1.6: Gram–Schmidt: $V_n \longrightarrow V_{n+1}$.

Note that

$$V_n \subset V_{n+1} \to \bigcup_n V_n \sim P_n \le P_{n+1} \to P$$

$$P_n^\perp \ge P_{n+1}^\perp \to P^\perp.$$

Assume $\bigcup_n V_n$ is dense in $\mathscr{H}$, then $P = I$ and $P^\perp = 0$. In lattice notations, we may write

$$\vee P_n = \sup P_n = I$$

$$\wedge P_n^\perp = \inf P_n^\perp = 0.$$

Lemma 1.40. *Let $P, Q \in Proj\,(\mathscr{H})$, then*

$$P + Q \in Proj(\mathscr{H}) \Longleftrightarrow PQ = QP = 0, \quad i.e., \ P \perp Q.$$

Proof. Notice that

$$(P + Q)^2 = P + Q + PQ + QP \tag{1.40}$$

and so

$$(P + Q)^2 = P + Q \tag{1.41}$$
$$\Updownarrow$$
$$PQ + QP = 0. \tag{1.42}$$

Suppose $PQ = QP = 0$ then

$$(P + Q)^2 = P + Q = (P + Q)^*, \quad i.e., \ P + Q \in Proj\,(\mathscr{H}).$$

Conversely, if $P + Q \in Proj(\mathscr{H})$, then $(P + Q)^2 = P + Q \Longrightarrow PQ + QP = 0$ by (1.42). Also, $(PQ)^* = Q^*P^* = QP$, combining with (1.42) yields

$$(PQ)^* = QP = -PQ. \tag{1.43}$$

Then,

$$(PQ)^2 = P(QP)Q \underset{(1.43)}{=} -P(PQ)Q = -PQ$$

which implies $PQ(I + PQ) = 0$. Hence,

$$PQ = 0 \quad \text{or} \quad PQ = I. \tag{1.44}$$

But by (1.43), PQ is skew-adjoint, it follows that $PQ = 0$, and so $QP = 0$.

$\square$

Lemma 1.41. *Let P and Q be two orthogonal projections (i.e., $P = P^* = P^2$) on a fixed Hilbert space $\mathscr{H}$, and consider the operator-norm $\|P + Q\|$. Suppose $\|P + Q\| \leq 1$; then the two projections are orthogonal, $PQ = 0$. As a result (Lemma 1.40), the sum $P + Q$ is itself an orthogonal projection.*

Proof. Let $u \in P\mathscr{H}$, i.e., $Pu = u$ holds. Then

$$\|u\|^2 \geq \|(P + Q)u\|^2 = \|u\|^2 + \|Qu\|^2 + \langle u, Qu \rangle + \langle Qu, u \rangle$$
$$= \|u\|^2 + 3\|Qu\|^2,$$

since $\langle Qu, u \rangle = \langle u, Qu \rangle = \|Qu\|^2$. Hence $Qu = 0$. Since

$$P\mathscr{H} = \{u \in \mathscr{H} : Pu = u\},$$

we get $QP = 0$; equivalently, $PQ = 0$; equivalently, $P\mathscr{H} \perp Q\mathscr{H}$. $\square$

The set of projections in a Hilbert space $\mathscr{H}$ is partially ordered according to the corresponding closed subspaces partially ordered by inclusion. Since containment implies commuting, the chain of projections

$$P_1 \leq P_2 \leq \cdots$$

is a family of commuting self-adjoint operators. By the spectral theorem, $\{P_i\}$ may be simultaneously diagonalized, so that P_i is unitarily equivalent to the operator of multiplication by χ_{E_i} on the Hilbert space $L^2(X, \mu)$, where X is compact and Hausdorff. Therefore the lattice structure of projections in $\mathscr{H}$ is precisely the lattice structure of χ_E, or equivalently, the lattice structure of measurable sets in X.

Lemma 1.42. *Consider $L^2(X, \mu)$. The following are equivalent.*

(i) $E \subset F$;
(ii) $\chi_E \chi_F = \chi_F \chi_E = \chi_E$;
(iii) $\|\chi_E f\| \leq \|\chi_F f\|$, *for any $f \in L^2$;*

(iv) $\chi_E \leq \chi_F$, *in the sense that*

$$\langle f, \chi_E f \rangle \leq \langle f, \chi_F f \rangle, \quad \forall f \in L^2(X).$$

Proof. The proof is trivial. Note that

$$\langle f, \chi_E f \rangle = \int \chi_E |f|^2 \, d\mu$$

$$\|\chi_E f\|^2 = \int |\chi_E f|^2 \, d\mu = \int \chi_E |f|^2 \, d\mu$$

where we used that fact that

$$\chi_E = \overline{\chi_E} = \chi_E^2.$$ □

What makes $Proj(\mathscr{H})$ intriguing is the non-commutativity. For example, if $P, Q \in Proj(\mathscr{H})$ are given, it does not follow (in general) that $P + Q \in Proj(\mathscr{H})$; nor that $PQP \in Proj(\mathscr{H})$. These two conclusions only hold if it is further assumed that P and Q commute; see Lemmas 1.40 and 1.43.

Lemma 1.43. *Let* $P, Q \in Proj(\mathscr{H})$; *then the following conditions are equivalent*:

(i) $PQP \in Proj(\mathscr{H})$;
(ii) $PQ = QP$.

Proof. First note that the operator $A = PQ - QP$ is skew-symmetric, i.e., $A^* = -A$, and so its spectrum is contained in the imaginary line $i\mathbb{R}$.

The implication (ii)⇒(i) above is immediate so assume (i), i.e., that

$$(PQP)^2 = PQP.$$

And using this, one checks by a direct computation that $A^3 = 0$. But with $A^* = -A$, and the spectral theorem, we therefore conclude that $A = 0$, in other words, (ii) holds. □

1.6 Unbounded operators between different Hilbert spaces

While the theory of unbounded operators has been focused on spectral theory where it is then natural to consider the setting of linear *endomorphisms* with dense domain in a fixed Hilbert space; many applications entail operators between distinct Hilbert spaces, say $\mathscr{H}_1$ and $\mathscr{H}_2$. Typically the facts given about the two differ greatly from one Hilbert space to the next.

Let $\mathcal{H}_i$, $i = 1, 2$ be two complex Hilbert spaces. The respective inner products will be written $\langle \cdot, \cdot \rangle_i$, with the subscript to identify the Hilbert space in question.

Definition 1.44. A linear operator T from $\mathcal{H}_1$ to $\mathcal{H}_2$ is a pair $(\mathcal{D}, T)$, where $\mathcal{D}$ is a linear subspace in $\mathcal{H}_1$, and $T\varphi \in \mathcal{H}_2$ is well-defined for all $\varphi \in \mathcal{D}$. We require linearity on $\mathcal{D}$; so

$$T(\varphi + c\psi) = T\varphi + cT\psi, \quad \forall \varphi, \psi \in \mathcal{D}, \ \forall c \in \mathbb{C}. \tag{1.45}$$

Notation. When T is given as in (1.45), we say that $\mathcal{D} = dom\,(T)$ is the domain of T, and

$$\mathcal{G}(T) = \left\{ \begin{pmatrix} \varphi \\ T\varphi \end{pmatrix}; \ \varphi \in \mathcal{D} \right\} \subset \begin{pmatrix} \mathcal{H}_1 \\ \oplus \\ \mathcal{H}_2 \end{pmatrix} \tag{1.46}$$

is the graph. By closure, we shall refer to closure in the norm of $\mathcal{H}_1 \oplus \mathcal{H}_2$, i.e.,

$$\left\| \begin{pmatrix} h_1 \\ h_2 \end{pmatrix} \right\|^2 = \|h_1\|_1^2 + \|h_2\|_2^2, \quad h_i \in \mathcal{H}_i. \tag{1.47}$$

If the closure $\overline{\mathcal{G}(T)}$ is the graph of a linear operator, we say that T is *closable*.

If $dom\,(T)$ is dense in $\mathcal{H}_1$, we say that T is densely defined. The abbreviated notation $\mathcal{H}_1 \xrightarrow{T} \mathcal{H}_2$ will be used when the domain of T is understood from the context.

Definition 1.45. Let $\mathcal{H}_1 \xrightarrow{T} \mathcal{H}_2$ be a densely defined operator, and consider the subspace $dom\,(T^*) \subset \mathcal{H}_2$ defined as follows:

$$dom\,(T^*) = \Big\{ h_2 \in \mathcal{H}_2; \ \exists C = C_{h_2} < \infty \ \text{s.t.}$$

$$|\langle T\varphi, h_2 \rangle_2| \leq C \|\varphi\|_1, \ \forall \varphi \in dom\,(T) \Big\}. \tag{1.48}$$

Then there is a unique $h_1 \in \mathcal{H}_1$ such that

$$\langle T\varphi, h_2 \rangle_2 = \langle \varphi, h_1 \rangle_1, \quad \text{and} \tag{1.49}$$

we set $T^* h_2 = h_1$.

It is immediate that T^* is an operator from $\mathscr{H}_2$ to $\mathscr{H}_1$; we write

$$
\begin{array}{ccc}
 & T & \\
\mathscr{H}_1 & \overrightarrow{} & \mathscr{H}_2 \\
 & \underleftarrow{} & \\
 & T^* &
\end{array}
\tag{1.50}
$$

The following holds:

Lemma 1.46. *Given a densely defined operator* $\mathscr{H}_1 \xrightarrow{T} \mathscr{H}_2$, *then* T *is closable if and only if* $\operatorname{dom}(T^*)$ *is dense in* $\mathscr{H}_2$.

Proof. See, e.g., [DS88]. Set $\chi : \mathscr{H}_1 \oplus \mathscr{H}_2 \longrightarrow \mathscr{H}_2 \oplus \mathscr{H}_1$,

$$
\chi \begin{pmatrix} h_1 \\ h_2 \end{pmatrix} = \begin{pmatrix} -h_2 \\ h_1 \end{pmatrix},
$$

then

$$
\mathscr{G}(T^*) = (\chi(\mathscr{G}(T)))^{\perp}.
\tag{1.51}
$$

For a general vector $\begin{pmatrix} h_2 \\ h_1 \end{pmatrix} \in \begin{pmatrix} \mathscr{H}_2 \\ \oplus \\ \mathscr{H}_1 \end{pmatrix}$ we have the following:

$$
\begin{pmatrix} h_2 \\ h_1 \end{pmatrix} \in (\chi(\mathscr{G}(T)))^{\perp}
$$

$$\Updownarrow$$

$$
\left\langle \begin{pmatrix} h_2 \\ h_1 \end{pmatrix}, \begin{pmatrix} -T\varphi \\ \varphi \end{pmatrix} \right\rangle = 0, \quad \forall \varphi \in \mathscr{D} \text{ (dense in } \mathscr{H}_1)
$$

$$\Updownarrow$$

$$
\langle h_2, T\varphi \rangle_2 = \langle h_1, \varphi \rangle_1, \quad \forall \varphi \in \mathscr{D}
$$

$$\Updownarrow$$

$$
h_2 \in \operatorname{dom}(T^*), \text{ and } T^* h_2 = h_1
$$

$$\Updownarrow$$

$$
\begin{pmatrix} h_2 \\ h_1 \end{pmatrix} \in \mathscr{G}(T^*).
$$

Notation. The symbol "$\perp$" denotes ortho-complement.

We must show that

$$\begin{pmatrix} 0 \\ h_2 \end{pmatrix} \in \mathscr{G}\left(T\right)^{\perp\perp} = \overline{\mathscr{G}\left(T\right)}$$

$$\updownarrow$$

$$h_2 \in dom\left(T^*\right)^{\perp} = \mathscr{H}_2 \ominus dom\left(T^*\right).$$

But this is immediate from

$$\mathscr{G}\left(T\right)^{\perp\perp} = \left(\chi\left(\mathscr{G}\left(T^*\right)\right)\right)^{\perp}. \tag{1.52}$$

$$\square$$

Remark 1.47 (Notation and facts).

(i) Let T be an operator $\mathscr{H}_1 \overset{T}{\to} \mathscr{H}_2$ and $\mathscr{H}_i$, $i = 1, 2$, two given Hilbert spaces. Assume $\mathscr{D} := dom\left(T\right)$ is dense in $\mathscr{H}_1$, and that T is *closable*. Then there is a unique *closed* operator, denoted $\overline{T}$, such that

$$\mathscr{G}\left(\overline{T}\right) = \overline{\mathscr{G}\left(T\right)} \tag{1.53}$$

where "—" on the RHS in (1.53) refers to norm closure in $\mathscr{H}_1 \oplus \mathscr{H}_2$, see (1.47).

(ii) We say that an operator $\mathscr{H}_1 \overset{T}{\to} \mathscr{H}_2$ is bounded if and only if $dom\left(T\right) = \mathscr{H}_1$, and $\exists C < \infty$ such that,

$$\|Th_1\|_2 \leq C \|h_1\|_1, \quad \forall h_1 \in \mathscr{H}_1, \tag{1.54}$$

where the subscripts refer to the respective Hilbert spaces. The infimum of the set of constants C in (1.54) is called the *norm* of T, written $\|T\|$.

(iii) If T is bounded, then so is T^*, in particular, $dom\left(T\right)^* = \mathscr{H}_2$, and the following identities hold:

$$\|T^*\| = \|T\|, \tag{1.55}$$

$$\|T^*T\| = \|T\|^2. \tag{1.56}$$

Note that formulas like (1.55) and (1.56) refer to the underlying Hilbert spaces, and that (1.55) writes out in more detail as follows:

$$\|T^*\|_{\mathscr{H}_2 \to \mathscr{H}_1} = \|T\|_{\mathscr{H}_1 \to \mathscr{H}_2}. \tag{1.57}$$

Example 1.48. An operator $T : \mathscr{H}_1 \longrightarrow \mathscr{H}_2$ with dense domain such that $dom\left(T^*\right) = 0$, i.e., "extremely" non-closable.

Let $J = [-\pi, \pi]$ be the 2π-period interval, $d\lambda_1 = dx$ the normalized Lebesgue measure, $\mu_3 =$ the middle-third Cantor measure. Recall the transformation of μ_3 is given by

$$\widehat{\mu}_3(\xi) = \int_{-\pi}^{\pi} e^{i\xi x} d\mu_3(x) = \prod_{n=1}^{\infty} \cos(\xi/3^n), \quad \xi \in \mathbb{R}. \tag{1.58}$$

Recall, the support E of μ_3 is the standard Cantor set, and that $\lambda_1(E) = 0$.

Now set $\mathscr{H}_1 = L^2(\lambda_1)$, and $\mathscr{H}_2 = L^2(\mu_3)$. The space $\mathscr{D} := C_c(-\pi, \pi)$ is dense in both $\mathscr{H}_1$ and in $\mathscr{H}_2$ with respect to the two L^2-norms. Hence, setting

$$T\varphi = \varphi, \quad \forall \varphi \in \mathscr{D}; \tag{1.59}$$

the identity operator, then becomes a Hilbert space operator $\mathscr{H}_1 \xrightarrow{T} \mathscr{H}_2$.

Using Definition 1.45, we see that $h_2 \in L^2(\mu_3)$ is in $dom(T^*)$ if and only if $\exists h_1 \in L^2(\lambda_1)$ such that

$$\int \varphi h_1 \, d\lambda_1 = \int \varphi h_2 \, d\mu_3, \quad \forall \varphi \in \mathscr{D}. \tag{1.60}$$

Since $\mathscr{D}$ is dense in both L^2-spaces, we get

$$\int_E h_1 \, d\lambda_1 = \int_E h_2 \, d\mu_3. \tag{1.61}$$

Now suppose $h_2 \neq 0$ in $L^2(\mu_3)$, then there is a subset $A \subset E$ such that $h_2 > 0$ on A, $\mu_3(A) > 0$, and $\int_A h_2 \, d\mu_3 > 0$. But $\int_A h_1 \, d\lambda_1 = \int_A h_2 \, d\mu_3$, and $\int_A h_1 \, d\lambda_1 = 0$ since $\lambda_1(A) = 0$. This contradiction proves that $dom(T^*) = 0$; and in particular T in (1.59) is unbounded and non-closable as claimed.

Remark 1.49. In Example 1.48, i.e., T is given by (1.59), and $\mathscr{H}_1 = L^2(\lambda_1)$, $\mathscr{H}_2 = L^2(\mu_3)$, $\lambda_1 =$ Lebesgue, $\mu_3 =$ Cantor measure, we have

$$\overline{\mathscr{G}(T)} = \begin{pmatrix} \mathscr{H}_1 \\ \oplus \\ \mathscr{H}_2 \end{pmatrix} = L^2(\lambda_1) \oplus L^2(\mu_3). \tag{1.62}$$

The proof of this fact (1.62) uses basic approximation properties of iterated function system (IFS) measures. Both λ_1 and μ_3 are IFS-measures. See, e.g., [Hut81, BHS05].

Definition 1.50. Let $\mathscr{H}_i$, $i = 1, 2$, be Hilbert spaces. A linear operator $J : \mathscr{H}_1 \to \mathscr{H}_2$ is said to be an *isometry* if

$$\|Jx\|_{\mathscr{H}_2} = \|x\|_{\mathscr{H}_1}, \tag{1.63}$$

for all $x \in \mathscr{H}_1$. Note that J is *not* assumed "onto."

J is said to be a *partial isometry* if (1.63) holds on $\mathscr{H}_1 \ominus ker(J)$.

Theorem 1.51 (The Polar Decomposition/Factorization). *Let* $\mathscr{H}_1 \xrightarrow{T} \mathscr{H}_2$ *be a densely defined operator, and assume that* $dom\,(T^*)$ *is dense in* $\mathscr{H}_2$, *i.e.,* T *is closable, then both of the operators* $T^*\overline{T}$ *and* $\overline{T}T^*$ *are densely defined, and both are self-adjoint.*

Moreover, there is a <u>*partial isometry*</u> $U : \mathscr{H}_1 \longrightarrow \mathscr{H}_2$ *with initial space in* $\mathscr{H}_1$ *and final space in* $\mathscr{H}_2$ *such that*

$$T = U\left(T^*\overline{T}\right)^{\frac{1}{2}} = \left(\overline{T}T^*\right)^{\frac{1}{2}} U. \tag{1.64}$$

(Equation (1.64) *is called the polar decomposition.)*

Proof. The proof of Theorem 1.51 will follow from the following more general considerations. See, e.g., [DS88]. □

Remark 1.52. If $\mathscr{H}_i$, $i = 1, 2, 3$, are three Hilbert spaces with operators $\mathscr{H}_1 \xrightarrow{A} \mathscr{H}_2 \xrightarrow{B} \mathscr{H}_3$, then the domain of BA is as follows:

$$dom\,(BA) = \{x \in dom\,(A)\,;\ Ax \in dom\,(B)\}\,,\quad \text{and} \tag{1.65}$$

on $x \in dom\,(BA)$, we have $(BA)\,(x) = B\,(Ax)$. In general, $dom\,(BA)$ may be 0, even if the two operators A and B have dense domains; see Example 1.48.

Definition 1.53 (The characteristic projection; see [Jør80, Sto51]). Let $\mathscr{H}_i$, $i = 1, 2$ be Hilbert spaces and $\mathscr{H}_1 \xrightarrow{T} \mathscr{H}_2$ a fixed operator; the characteristic projection P of T is the projection in $\mathscr{H}_1 \oplus \mathscr{H}_2$ onto $\overline{\mathscr{G}\,(T)}$. We shall write

$$P = \begin{pmatrix} P_{11} & P_{12} \\ P_{21} & P_{22} \end{pmatrix}$$

where the components are bounded operators as follows:

$$P_{11} : \mathscr{H}_1 \longrightarrow \mathscr{H}_1,\ P_{12} : \mathscr{H}_2 \longrightarrow \mathscr{H}_1,$$

$$P_{21} : \mathscr{H}_1 \longrightarrow \mathscr{H}_2,\ P_{22} : \mathscr{H}_2 \longrightarrow \mathscr{H}_2.$$

Since

$$P = P^* = P^2 \tag{1.66}$$

we have

$$P_{11} = P_{11}^* \geq 0,\qquad P_{12} = P_{21}^*,$$

$$P_{21} = P_{12}^*,\ \text{and}\quad P_{22} = P_{22}^* \geq 0,$$

where "$\geq$" refers to the natural order as self-adjoint operators.

From (1.66), we further get

$$P_{ij} = \sum_{k=1}^{2} P_{ik} P_{kj}, \quad \forall i, j = 1, 2. \tag{1.67}$$

Lemma 1.54. *Let T be as above, and let $P = P(T) = (P_{i,j})_{i,j=1}^{2}$ be its characteristic projection, then T is closable if and only if*

$$\ker \left(I_{\mathscr{H}_2} - P_{22} \right) = 0; \tag{1.68}$$

i.e., if and only if the following implication holds:

$$\boxed{y \in \mathscr{H}_2, \ P_{22} y = y} \Longrightarrow \boxed{y = 0} \tag{1.69}$$

Proof. If $P = (P_{ij})$ is the characteristic projection for T, it follows from (1.51) and Lemma 1.46, that

$$P^{\vee} := \begin{pmatrix} I_2 - P_{22} & P_{21} \\ P_{12} & I_1 - P_{11} \end{pmatrix} \text{ in } \begin{pmatrix} \mathscr{H}_2 \\ \oplus \\ \mathscr{H}_1 \end{pmatrix} \tag{1.70}$$

is the characteristic projection matrix for the operator $T^* : \mathscr{H}_2 \longrightarrow \mathscr{H}_1$.
    Hence

$$T^* \left(y - P_{22} y \right) = P_{12} y, \quad \forall y \in \mathscr{H}_2, \text{ and} \tag{1.71}$$

$$T^* P_{21} x = x - P_{11} x, \quad \forall x \in \mathscr{H}_1. \tag{1.72}$$

Note that T is closable if and only if

$$\begin{pmatrix} 0 \\ y \end{pmatrix} \in \overline{\mathscr{G}(T)} \Longrightarrow y = 0. \tag{1.73}$$

But by (1.67), $\begin{pmatrix} 0 \\ y \end{pmatrix} \in \overline{\mathscr{G}(T)}$ holds if and only if $P_{12} y = 0$ and $P_{22} y = y$.

    But by (1.71), we have $P_{22} y = y \Longrightarrow P_{12} y = 0$, and the desired conclusion follows. $\qquad\square$

We now turn to the proof of Theorem 1.51. Given is a closable operator $\mathscr{H}_1 \xrightarrow{T} \mathscr{H}_2$ with dense domain. To simplify notation, we shall denote $\overline{T}$ also by T.

Proof. (Theorem 1.51) By passing to the closure $\overline{T}$ of the given operator T, we get $\overline{\mathscr{G}(T)} = \mathscr{G}(\overline{T})$, where $\overline{T}$ is then a closed operator. It is easy to see that

$$\left(\overline{T} \right)^* = T^* \tag{1.74}$$

on its dense domain in $\mathcal{H}_2$. Below, we shall write T for the associated closed operator $\overline{T}$.

In Lemma 1.54 we formed the characteristic projection for both T and T^*; see especially (1.70). As a result we get

$$\begin{cases} T P_{11} = P_{21} & \text{on } \mathcal{H}_1, \text{ and} \\ T P_{12} = P_{22} & \text{on } \mathcal{H}_2. \end{cases} \tag{1.75}$$

Similarly (see (1.70)),

$$\begin{cases} T^* (I - P_{22}) = P_{12} & \text{on } \mathcal{H}_2, \text{ and} \\ T^* P_{21} = I - P_{11} & \text{on } \mathcal{H}_1. \end{cases} \tag{1.76}$$

Note that part of the conclusion in (1.75) is that $ran(P_{11}) = \{P_{11}x; x \in \mathcal{H}_1\}$ is contained in $dom(T)$.

Further note that

$$\ker(P_{11}) = ran(P_{11})^{\perp} = 0, \quad \text{and} \tag{1.77}$$

$$\ker(I - P_{22}) = (ran(I - P_{22}))^{\perp} = 0. \tag{1.78}$$

We already proved (1.78), and we now turn to (1.77).

Combining (1.75) and (1.76), we get $T^*T P_{11} = I - P_{11}$, and so

$$(1 + T^*T) P_{11} = I; \tag{1.79}$$

equivalently,

$$P_{11}x + T^*T P_{11}x = x, \quad \forall x \in \mathcal{H}_1.$$

So $ran(P_{11})$ is dense in $\mathcal{H}_1$, and $ran(P_{11}) \subset dom(T^*T)$; so $dom(T^*T)$ is dense in $\mathcal{H}_1$. By (1.79) therefore,

$$T^*T = (P_{11})^{-1} - I \tag{1.80}$$

is a self-adjoint operator with dense domain $ran(P_{11})$ in $\mathcal{H}_1$.

The conclusions in Theorem 1.51 now follow. $\qquad\square$

Corollary 1.55. *Let $\mathcal{H}_1 \xrightarrow{T} \mathcal{H}_2$ be a closed operator with dense domain in $\mathcal{H}_1$, and let T^*T and TT^* be the self-adjoint operators from Theorem 1.51. Then the entries of the characteristic matrix are as follows:*

$$\begin{pmatrix} P_{11} = (I_1 + T^*T)^{-1} & P_{12} = T^* (I_2 + TT^*)^{-1} \\ P_{21} = T (I_1 + T^*T)^{-1} & P_{22} = I_2 - (I_2 + TT^*)^{-1} \end{pmatrix} \tag{1.81}$$

where I_i denote the identity operators in the respective Hilbert spaces, $i = 1, 2$.

Corollary 1.56. *Let $\mathscr{H}_1 \xrightarrow{T} \mathscr{H}_2$ be as above (T densely defined but not assumed closable), and let $(P_{ij})_{ij=1}^2$ be the characteristic projection referring to the sum Hilbert space $\mathscr{H}_1 \oplus \mathscr{H}_2$. Let Q = the projection onto $(I_2 - P_{22})\,\mathscr{H}_2 = (\ker(I_2 - P_{22}))^\perp$, then*

$$
\begin{array}{cc}
 & \begin{array}{cc} \mathscr{H}_1 & Q\mathscr{H}_2 \end{array} \\
\begin{array}{c} \mathscr{H}_1 \\ Q\mathscr{H}_2 \end{array} & \left(\begin{array}{c|c} P_{11} & P_{12}Q \\ \hline QP_{21} & P_{22}Q \end{array} \right)
\end{array}
\tag{1.82}
$$

is a characteristic projection of a closable operator T_{clo}, now referring to the sum Hilbert space $\mathscr{H}_1 \oplus Q\mathscr{H}_2$.

Moreover, T_{clo} is given by

$$
T_{clo}x = \lim_{n \to \infty} \frac{1}{n+1} \sum_{k=1}^n k\, P_{22}^{n-k} P_{21}x,
\tag{1.83}
$$

with dom(T_{clo}) = dom(T) dense in $\mathscr{H}_1$.

Proof. (Sketch) Let T, (P_{ij}), Q and x be as given above; then

$$
\frac{1}{n+1} \sum_{k=0}^n P_{22}^k Tx = Tx - \frac{1}{n+1} \sum_{k=1}^n k\, P_{22}^{n-k} P_{21}x.
\tag{1.84}
$$

Note that the LHS in (1.84) converges in the norm of $\mathscr{H}_1$ by virtue of the ergodic theorem (see [Yos95]) applied to the self-adjoint contraction P_{22}. Moreover, the limit is $Q^\perp Tx$ where $Q^\perp := I - Q = \mathrm{proj}\,(\ker(I_2 - P_{22}))$. $\square$

1.7 Normal operators

The reader may already be familiar some version or the other of the *Spectral Theorem*. Below we shall present it in its general form via projection valued measures, and the corresponding Hilbert space-direct integral decomposition. To suggest an analogy to the finite-dimensional case (matrices), we shall refer to the direct integral representation for an operator as a "diagonalization." The most general class of operators admitting this kind of "diagonalization" is the class of *normal operators*.

We shall include unbounded normal operators (important for our applications to *representation theory*). In the unbounded case, one must make a distinction between "formally normal" and normal. Only the latter allows a "diagonalization". As is always the case for unbounded operators, domains will be important; see Theorems 1.57 and 1.58. Naturally, the class of normal operators includes the self-adjoint operators, and the unitary operators. An important approach to the study of a given unbounded operator, say A is consideration of the projection onto the graph of A; see Corollary 1.55 and Theorem 1.59. Properties of this projection further yields a practical criterion for deciding when a formally normal operator is indeed normal, in the sense of admitting a "diagonalization" via a projection valued measure.

The spectral theorem for strongly *continuous one-parameter groups of unitary operators* is especially relevant for understanding dynamics in quantum physics, and it goes by the name *Stone's Theorem*, Theorem 1.66, see Section 1.9.

We now return to the general setting, and resume our study the axiomatic case of densely defined unbounded closable operators acting in a fixed Hilbert space $\mathscr{H}$. Our aim is to give necessary and sufficient conditions for when the given operator A is normal, as an operator in $\mathscr{H}$. In Theorem 1.59, we give a set of necessary and sufficient conditions for A to be normal, and expressed in terms of the associated characteristic matrix. Since the initial operator A is unbounded, there are subtle issues having to do with domains. We first address these issues in Theorems 1.57 and 1.58.

Theorem 1.57 below concerning operators of the form A^*A is an application of Stone's characteristic matrix.

Theorem 1.57 (von Neumann). *If A is a regular operator in a Hilbert space $\mathscr{H}$, then*

(i) A^*A *is self-adjoint;*
(ii) $\mathscr{D}(A^*A)$ *is a core of A, i.e.,*

$$\overline{A\big|_{\mathscr{D}(A^*A)}} = A;$$

(iii) *In particular, $\mathscr{D}(A^*A)$ is dense in $\mathscr{H}$.*

Proof. Let (P_{ij}) be the characteristic matrix of A (Definition 1.53). Note that $A^*A = P_{11}^{-1} - 1$. Since P_{11} is self-adjoint, so is P_{11}^{-1}. Thus, AA^* is self-adjoint.

Suppose $(a, Aa) \in \mathscr{G}(A)$ such that

$$(a, Aa) \perp \mathscr{G}(A\big|_{\mathscr{D}(A^*A)}); \text{ i.e.,}$$

$$\langle a, b \rangle + \langle Aa, Ab \rangle = \langle a, (1 + A^*A) b \rangle = 0, \quad \forall b \in \mathscr{D}(A^*A).$$

Since $1 + A^*A = P_{11}^{-1}$, and P_{11} is a bounded operator, then

$$\mathscr{R}(1 + A^*A) = \mathscr{D}(P_{11}) = \mathscr{H}.$$

It follows that $a \perp \mathscr{H}$, and so $a = 0$. □

Theorem 1.58 (von Neumann). *Let A be a regular operator in a Hilbert space $\mathscr{H}$. Then A is normal if and only if $\mathscr{D}(A) = \mathscr{D}(A^*)$ and $\|Aa\| = \|A^*a\|$, for all $a \in \mathscr{D}(A)$.*

Proof. Suppose A is normal. Then for all $a \in \mathscr{D}(A^*A) (= \mathscr{D}(AA^*))$, we have

$$\|Aa\|^2 = \langle Aa, Aa \rangle = \langle a, A^*Aa \rangle = \langle a, AA^*a \rangle = \langle Aa, A^*a \rangle = \|A^*a\|^2 ;$$

i.e., $\|Aa\| = \|A^*a\|$, for all $a \in \mathscr{D}(A^*A)$. It follows that

$$\mathscr{D}\left(\overline{A\big|_{\mathscr{D}(A^*A)}}\right) = \mathscr{D}\left(\overline{A^*\big|_{\mathscr{D}(AA^*)}}\right).$$

By Theorem 1.57, $\mathscr{D}(A) = \mathscr{D}(\overline{A\big|_{\mathscr{D}(A^*A)}})$ and $\mathscr{D}(A^*) = \mathscr{D}(\overline{A^*\big|_{\mathscr{D}(AA^*)}})$. Therefore, $\mathscr{D}(A) = \mathscr{D}(A^*)$ and $\|Aa\| = \|A^*a\|$, for all $a \in \mathscr{D}(A)$.

Conversely, the map $Aa \mapsto A^*a$, $a \in \mathscr{D}(A)$, extends uniquely to a partial isometry V with initial space $\overline{\mathscr{R}(A)}$ and final space $\overline{\mathscr{R}(A^*)}$, such that $A^* = VA$ and so $A = A^*V^*$. Then $A^*A = A^*(V^*V)A = (A^*V^*)(VA) = AA^*$. Thus, A is normal. □

The following theorem is due to M.H. Stone (see [Sto90], first ed. 1932, and [Sto51]).

Theorem 1.59. *Let A be a regular operator in a Hilbert space $\mathscr{H}$. Let $P = (P_{ij})$ be the characteristic matrix of A. The following are equivalent.*

(i) *A is normal.*
(ii) *P_{ij} are mutually commuting.*
(iii) *A is affiliated with an Abelian von Neumann algebra.*

Proof. Assuming (i)⇔(ii), we prove that (i)⇔(iii).

Suppose A is normal, i.e. P_{ij} are mutually commuting. Then A is affiliated with the Abelian von Neumann algebra $\{P_{ij}\}''$. For if $B \in \{P_{ij}\}'$, then B commutes P_{ij}, and so B commutes with A.

Conversely, if A is affiliated with an Abelian von Neumann algebra $\mathfrak{M}$, then $P_{ij} \in \mathfrak{M}$. This shows that P_{ij} are mutually commuting, and A is normal. □

1.8　Closable pairs of operators

A key role in analysis is based on notions of duality, and our introduction below of *closable pairs* is a case in point. In its simplest form, it occurs in consideration of integration of parts, and in variety of boundary value problems from PDE theory. Hence, the setting of *unbounded operators* is essential. In considering pairs of Hilbert spaces, say $\mathscr{H}_i$, $i = 1, 2$, and general duality theory, this particular duality, referring to pairs of operators, the two operators are as follows; one is acting from $\mathscr{H}_1$ to $\mathscr{H}_2$, and the other in the opposite direction, from $\mathscr{H}_2$ to $\mathscr{H}_1$. Both operators will be assumed to have dense domains in their respective Hilbert spaces. We shall say that two such operators form a *closable pair* if the condition in Definition 1.60 is satisfied. The relevant literature is extensive, and we point the reader to the discussion of the literature at the end of the chapter (Section 1.10), but for fundamentals of unbounded operator theory, see, e.g., [DS88].

Definition 1.60 (Closable pairs). Let $\mathscr{H}_1$ and $\mathscr{H}_2$ be two Hilbert spaces with respective inner products $\langle \cdot, \cdot \rangle_i$, $i = 1, 2$; let $\mathscr{D}_i \subset \mathscr{H}_i$, $i = 1, 2$, be two dense linear subspaces; and let

$$S_0 : \mathscr{D}_1 \to \mathscr{H}_2, \quad \text{and} \quad T_0 : \mathscr{D}_2 \to \mathscr{H}_1$$

be linear operators such that

$$\langle S_0 u, v \rangle_2 = \langle u, T_0 v \rangle_1, \quad \forall u \in \mathscr{D}_1, \forall v \in \mathscr{D}_2. \tag{1.85}$$

Then both operators S_0 and T_0 are closable. The closures $S = \overline{S_0}$, and $T = \overline{T_0}$ satisfy

$$S \subseteq T^* \quad \text{and} \quad T \subseteq S^*. \tag{1.86}$$

We say the system (S_0, T_0) is a *closable pair*. (Also see Definition 5.5.)

Example 1.61. An operator $T : \mathscr{H}_1 \longrightarrow \mathscr{H}_2$ with dense domain such that $dom\,(T^*) = 0$, i.e., "extremely" non-closable.

Set $\mathscr{H}_i = L^2(\mu_i)$, $i = 1, 2$, where μ_1 and μ_2 are two mutually singular measures on a fixed locally compact measurable space, say X. The space $\mathscr{D} := C_c(X)$ is dense in both $\mathscr{H}_1$ and in $\mathscr{H}_2$ with respect to the two L^2-norms. Then, the identity mapping $T\varphi = \varphi$, $\forall \varphi \in \mathscr{D}$, becomes a Hilbert space operator $\mathscr{H}_1 \xrightarrow{T} \mathscr{H}_2$.

We see that $h_2 \in L^2(\mu_2)$ is in $dom\,(T^*)$ if and only if $\exists h_1 \in L^2(\mu_1)$ such that

$$\int \varphi h_1 \, d\mu_1 = \int \varphi h_2 \, d\mu_2, \quad \forall \varphi \in \mathscr{D}. \tag{1.87}$$

Since $\mathscr{D}$ is dense in both L^2-spaces, we get

$$\int_E h_1 \, d\mu_1 = \int_E h_2 \, d\mu_2, \tag{1.88}$$

where $E = supp\,(\mu_2)$.

Now suppose $h_2 \neq 0$ in $L^2\,(\mu_2)$, then there is a subset $A \subset E$ such that $h_2 > 0$ on A, $\mu_2\,(A) > 0$, and $\int_A h_2 \, d\mu_2 > 0$. But $\int_A h_1 \, d\mu_1 = \int_A h_2 \, d\mu_2$, and $\int_A h_1 \, d\mu_1 = 0$ since $\mu_1\,(A) = 0$. This contradiction proves that $dom\,(T^*) = 0$; and in particular T is unbounded and non-closable.

Below we include a structure theorem characterizing duality for closable pairs of operators.

Theorem 1.62. *Let $\mathscr{H}_i$ be Hilbert spaces with inner products $\langle \cdot, \cdot \rangle_i$, $i = 1, 2$. Let $\mathscr{D}$ be a vector space s.t. $\mathscr{D} \subset \mathscr{H}_1 \cap \mathscr{H}_2$, and suppose*

$$\mathscr{D} \text{ is dense in } \mathscr{H}_1. \tag{1.89}$$

Set $\mathscr{D}^* \subset \mathscr{H}_2$,

$$\mathscr{D}^* = \{h \in \mathscr{H}_2 \mid \exists C_h < \infty \text{ s.t. } |\langle \varphi, h \rangle_2| \leq C_h \, \|\varphi\|_1 , \; \forall \varphi \in \mathscr{D}\}; \tag{1.90}$$

then the following two conditions (i)–(ii) are equivalent:

(i) $\mathscr{D}^*$ *is dense in $\mathscr{H}_2$; and*
(ii) *there is a self-adjoint operator Δ with dense domain in $\mathscr{H}_1$ s.t. $\mathscr{D} \subset dom\,(\Delta)$, and*

$$\langle \varphi, \Delta\varphi \rangle_1 = \|\varphi\|_2^2 , \quad \forall \varphi \in \mathscr{D}. \tag{1.91}$$

Proof. (i)$\Longrightarrow$(ii) Assume $\mathscr{D}^*$ is dense in $\mathscr{H}_2$; then by (1.90), the inclusion operator

$$J : \mathscr{H}_1 \longrightarrow \mathscr{H}_2, \quad J\varphi = \varphi, \; \forall \varphi \in \mathscr{D} \tag{1.92}$$

has $\mathscr{D}^* \subset dom(J^*)$; so by (i), J^* has dense domain in $\mathscr{H}_2$, and J is closable. By von Neumann's theorem (see Theorem 1.57), $\Delta := J^*\overline{J}$ is self-adjoint in $\mathscr{H}_1$; clearly $\mathscr{D} \subset dom\,(\Delta)$; and for $\varphi \in \mathscr{D}$,

$$\text{LHS}_{(1.91)} = \langle \varphi, J^* J\varphi \rangle_1 = \langle J\varphi, J\varphi \rangle_2 \underset{\text{by (1.92)}}{=} \|\varphi\|_2^2 = \text{RHS}_{(1.91)}.$$

(Note that $J^{**} = \overline{J}$.)

Claim 1.63. $\mathscr{D}^* = dom(J^*)$, $\mathscr{D} \subset \mathscr{H}_1 \overset{J}{\underset{J^*}{\rightleftarrows}} \mathscr{H}_2 \supset \mathscr{D}^*$.

Proof. $h \in dom(J^*) \iff \exists C = C_h < \infty$ s.t.

$$|\langle \underbrace{J\varphi}_{=\varphi}, h \rangle_2| \leq C \|\varphi\|_1, \quad \forall \varphi \in \mathscr{D} \iff h \in \mathscr{D}^*, \text{ by Definition } (1.90).$$

Since $dom(J^*)$ is dense, J is closable, and by von Neumann's theorem $\Delta := J^* \overline{J}$ is self-adjoint in $\mathscr{H}_1$. $\qquad\square$

(ii)$\implies$(i) Assume (ii); then we get a well-defined partial isometry K : $\mathscr{H}_1 \longrightarrow \mathscr{H}_2$, by

$$K\Delta^{\frac{1}{2}}\varphi = \varphi, \quad \forall \varphi \in \mathscr{D}. \tag{1.93}$$

Indeed, (1.91) reads:

$$\|\Delta^{\frac{1}{2}}\varphi\|_1^2 = \langle \varphi, \Delta\varphi \rangle_1 = \|\varphi\|_2^2, \quad \varphi \in \mathscr{D},$$

which means that K in (1.93) is a *partial isometry* with $dom(K) = K^*K = \overline{ran(\Delta^{\frac{1}{2}})}$; and we set $K = 0$ on the complement in $\mathscr{H}_1$.

Then the following inclusion holds:

$$\left\{ h \in \mathscr{H}_2 \mid K^*h \in dom(\Delta^{\frac{1}{2}}) \right\} \subseteq \mathscr{D}^*. \tag{1.94}$$

We claim that LHS in (1.94) is dense in $\mathscr{H}_2$; and so (i) is satisfied. To see that (1.94) holds, suppose $K^*h \in dom(\Delta^{\frac{1}{2}})$; then for all $\varphi \in \mathscr{D}$, we have

$$|\langle h, \varphi \rangle_2| = \left| \langle h, K\Delta^{\frac{1}{2}}\varphi \rangle_2 \right|$$

$$= \left| \langle K^*h, \Delta^{\frac{1}{2}}\varphi \rangle_1 \right| \quad \text{(by (1.93))}$$

$$= \left| \langle \Delta^{\frac{1}{2}}K^*h, \varphi \rangle_1 \right| \leq \|\Delta^{\frac{1}{2}}K^*h\|_1 \|\varphi\|_1,$$

where we used Schwarz for $\langle \cdot, \cdot \rangle_1$ in the last step. $\qquad\square$

Corollary 1.64. *Let $\mathscr{D} \subset \mathscr{H}_1 \cap \mathscr{H}_2$ be as in the statement of Theorem 1.62, and let $J : \mathscr{H}_1 \longrightarrow \mathscr{H}_2$ be the associated closable operator; see (1.92). Then the complement*

$$\mathscr{H}_2 \ominus \mathscr{D} = \{ h \in \mathscr{H}_2 \mid \langle \varphi, h \rangle_2 = 0, \ \forall \varphi \in \mathscr{D} \}$$

satisfies $\mathscr{H}_2 \ominus \mathscr{D} = ker(J^*)$.

Proof. Immediate from the theorem. $\qquad\square$

1.9 Stone's Theorem

The gist of the result (Theorem 1.66) is as follows: Given a fixed Hilbert space, there is then a 1-1 correspondence between any two in pairs from the following three: (i) strongly continuous unitary one-parameter groups $\mathcal{U}(t)$; (ii) self-adjoint operators H (generally unbounded) with dense domain; and (iii) projection valued measures $P(\cdot)$, abbreviated PVM. Starting with $\mathcal{U}(t)$, we say that the corresponding self-adjoint operator H is its generator, and then the PVM $P(\cdot)$ will be from the Spectral Theorem applied to H.

Definition 1.65. A unitary one-parameter group is a function $\mathcal{U} : \mathbb{R} \to$ (unitary operators in $\mathscr{H}$), such that

$$\mathcal{U}(s+t) = \mathcal{U}(s)\mathcal{U}(t), \quad \forall s, t \in \mathbb{R}; \tag{1.95}$$

and for $\forall h \in \mathscr{H}$,

$$\lim_{t \to 0} \mathcal{U}(t) h = h \text{ (strong continuity)}. \tag{1.96}$$

Theorem 1.66 (Stone's Theorem [Sto90], first ed. 1932). *There is a sequence of bijective correspondences between* (i)–(iii) *below, i.e.,* (i) $\Rightarrow$ (ii) $\Rightarrow$ (iii) $\Rightarrow$ (i):

(i) *PVMs $P(\cdot)$ (see Definition 1.35);*
(ii) *unitary one-parameter groups $\mathcal{U}$; and*
(iii) *self-adjoint operators H with dense domain in $\mathscr{H}$.*

Proof. The correspondence is given explicitly as follows:
(i)$\Rightarrow$ (ii): Given P, a PVM, set

$$\mathcal{U}(t) = \int_{\mathbb{R}} e^{i\lambda t} P(d\lambda) \tag{1.97}$$

where the integral on the RHS in (1.97) is the limit of finite sums of

$$\sum_k e^{i\lambda_k t} P(E_k), \quad t \in \mathbb{R}; \tag{1.98}$$

$E_i \cap E_j = \emptyset$ $(i \neq j)$, $\bigcup_k E_k = \mathbb{R}$.
(ii) $\Rightarrow$ (iii): Given $\{\mathcal{U}(t)\}_{t \in \mathbb{R}}$, set

$$dom(H) = \left\{ f \in \mathscr{H}, \text{ s.t. } \lim_{t \to 0_+} \frac{1}{it} (\mathcal{U}(t) f - f) \text{ exists} \right\}$$

and

$$iHf = \lim_{t \to 0_+} \frac{\mathcal{U}(t) f - f}{t}, \quad f \in dom(H), \tag{1.99}$$

then $H^* = H$.

(iii) $\Rightarrow$ (i): Given a self-adjoint operator H with dense domain in $\mathscr{H}$; then by the spectral theorem (Section 1.7) there is a unique PVM, $P(\cdot)$ such that

$$H = \int_{\mathbb{R}} \lambda P(d\lambda); \quad \text{and} \tag{1.100}$$

$$\square$$

$$dom(H) = \left\{ f \in \mathscr{H}; \text{ s.t. } \int_{\mathbb{R}} \lambda^2 \|P(d\lambda) f\|^2 < \infty \right\}. \tag{1.101}$$

Also see [Lax02, RS75, Rud91] for more details.

Remark 1.67. Note that the selfadjointness condition on H in (iii) in Theorem 1.66 is stronger than merely Hermitian symmetry, i.e., the condition

$$\langle Hu, v \rangle = \langle u, Hv \rangle \tag{1.102}$$

for all pairs of vectors u and $v \in dom(H)$. We shall discuss this important issue in much detail in this book, both in connection with the theory, and its applications. The applications are in physics, statistics, and infinite networks.

Observations. Introducing the adjoint operator H^*, we note that (1.102) is equivalent to

$$H \subset H^*, \quad \text{or} \tag{1.103}$$

$$\mathscr{G}(H) \subset \mathscr{G}(H^*), \tag{1.104}$$

where $\mathscr{G}$ denotes the graph of the respective operators and where (1.103) & (1.104) mean that $dom(H) \subset dom(H^*)$, and $Hu = H^*u$ for $\forall u \in dom(H)$.

If (1.103) holds, then H may, or may not, have *self-adjoint extensions*.

Definition 1.68. Let H be a densely defined, closed, Hermitian operator in $\mathscr{H}$. The closed subspaces

$$\mathscr{D}_\pm(H) = ker(H^* \mp i)$$

$$= \{\xi \in \mathscr{D}(H^*) : H^*\xi = \pm i\,\xi\}$$

are called the *deficiency spaces* of H. We introduce two indices $d_\pm$ (deficiency-indices)

$$d_\pm = \dim \mathscr{D}_\pm(H). \tag{1.105}$$

The following characterization of self-adjoint extensions is due to von Neumann:

Theorem 1.69 (von Neumann).

(i) *Suppose $H \subset H^*$, then H has self-adjoint extensions if and only if $d_+ = d_-$.*

(ii) *If H has self-adjoint extensions, say K (i.e., $K^* = K$,) so $H \subset K$, then it follows that*

$$H \subset K \subset H^*. \tag{1.106}$$

So, if there are self-adjoint extensions, they lie between H and H^*.

(iii) *The (closed) Hermitian extensions of H are indexed by partial isometries with initial space in $\mathscr{D}_+(H)$ and final space in $\mathscr{D}_-(H)$. Given a partial isometry U, the Hermitian extension $\widetilde{H_U} \supset H$ is determined as follows:*

$$\widetilde{H}_U \left(x + (1+U)\, x_+\right) = Hx + i\,(1-U)\, x_+, \quad \text{where} \tag{1.107}$$

$$\mathscr{D}\left(\widetilde{H}_U\right) = \{x + x_+ + Ux_+ : x \in \mathscr{D}(H),$$

$$x_+ \in \mathscr{D}_+(H)\}. \tag{1.108}$$

Proof. See, e.g., [DS88, JT17]. □

Definition 1.70. If $H \subset H^*$, and if the closure $\overline{H} = H^{**}$ is self-adjoint, we say that H is *essentially self-adjoint*.

Remark 1.71 (The Friedrichs extension (named after Kurt Friedrichs)). While we have omitted a detailed discussion, we refer to the cited literature, especially [DS88]. In brief outline, starting with a positive semidefinite (non-negative), densely defined symmetric operator T in a fixed Hilbert space, say $\mathscr{H}$, the Friedrichs extension is a canonical self-adjoint extension of a given non-negative densely defined symmetric operator T. It is "canonical" because it arises as a natural Hilbert completion, arising from T. But the verification of its properties, is a bit tricky; omitted here.

Note that the initial semibounded operator T may fail to be essentially self-adjoint, it may have non-zero deficiency indices (n, n). The Friedrichs extension is self-adjoint and is also semibounded, with the same lower bound 0. Further, note that there are other self-adjoint extension with the same lower bound. In fact there is an order relation on such semibounded self-adjoint extension of T, with the order admitting a minimal (Friedrichs), and maximal (Krein) operator.

1.10 Guide to the literature

While we have cited some paper from the Bibliography in our discussion inside the chapter, readers not familiar with the ideas involved may wish to consult the following sources: [Aro50, BR79, BR81, DS88, Jan97, JT17, Lax02, Leb05, Nel69, RS75, Rud91, Sto51, vN32a, Yos95, Hid80].

Chapter 2

Infinite-Dimensional Algebraic Systems: Lie Algebras, Algebras with Involution (∗-Algebras), and the Canonical Commutation Relations (CCRs)

> *Representation theory is a branch of mathematics that studies abstract algebraic structures by representing their elements as linear transformations of vector spaces, and studies modules over these abstract algebraic structures.* John von Neumann.

While the subject of infinite-dimensional algebraic structures is vast, with many and varied subdisciplines, and with many and diverse applications, for our present purposes we have made choices as follows, each motivated by the aim of the book: The CCR-algebra, and the Fock representations, symmetric Fock space (Chapter 3), Malliavin calculus (Chapter 5), and Gaussian white noise process (see, e.g., Sections 3.3, 4.1.2 and 4.4). Useful supplement treatments in the literature: [Aro50, AW73, Kol83].

The discussion below is motivated by applications, especially the features which are dictated by quantum physics, the case of quantum fields. In quantum physics, particles (or waves) force two kinds of non-commutativity. They go by the names, the canonical commutation relations (CCRs), and the canonical anti-commutation relations (CAR). They will be discussed in detail below, but we stress that the study of non-commutative systems has a host of applications in diverse areas of mathematics and applications, in addition to those from physics.

We study densely defined unbounded operators acting between different Hilbert spaces. For these, we introduce a notion of symmetric (closable) pairs of operators. The purpose is to give applications to selected themes at the cross road of operator commutation relations and stochastic calculus. We study a family of representations of the canonical commutation relations (CCR)-algebra (an infinite number of degrees of freedom), which we call admissible. The family of admissible representations includes the Fock-vacuum representation. We show that, to every admissible representation, there is an associated Gaussian stochastic calculus, and we point out that the case of the Fock-vacuum CCR-representation in a natural way yields the operators of Malliavin calculus. We thus get the operators of Malliavin's calculus of variation from a more algebraic approach than is common. We further obtain explicit and natural formulas, and rules, for the operators of stochastic calculus. Our approach makes use of a notion of symmetric (closable) pairs of operators. The Fock-vacuum representation yields a maximal symmetric pair. This duality viewpoint has the further advantage that issues with unbounded operators and dense domains can be resolved much easier than what is possible with alternative tools. With the use of CCR representation theory, we also obtain, as a byproduct, a number of new results in multivariable operator theory which we feel are of independent interest.

In mathematical physics, there are two important (called canonical) dual pairs of observables, or fields; one based on a relation for commutators, and the other, for anti-commutators. In each case, we get algebras of observables, or fields; they are abbreviated as follows: the CCR algebras (for canonical commutation relations), and CAR algebras (for canonical anticommutation relations). Which to use is dictated by quantum physics: bosons (named after Bose) have statistics governed by the CCRs, while fermions (named after Fermi) have statistics governed by the CARs. Each case plays a fundamental role in both quantum statistical mechanics, and in quantum field theory. For additional details, covering both the basics of the mathematics and physics, see, e.g., [BR81, GJ87, Ols20]. In both cases, CCR and CAR, a Hilbert space $\mathscr{H}$ is specified at the outset, and the generators for the respective algebras are chosen to be vectors (states) in $\mathscr{H}$. The respective algebras are then $CCR(\mathscr{H})$, and $CAR(\mathscr{H})$. The study of representations of the first one is of special relevance to our present aim, stochastic analysis. Representations of CCR($\mathcal{H}$) entails unbounded operators, while the second algebra CAR($\mathcal{H}$) is a C^*-algebra, and bounded operators suffice for its representations. Our choice of CCR($\mathcal{H}$), and its representations, is dictated by our applications to infinite-dimensional calculus for Gaussian processes.

In both instances, we shall be interested in the case when the initial Hilbert space $\mathcal{H}$ is infinite-dimensional, for example $\mathcal{H} = L^2(\mu)$ for a suitable choice of measure μ. Moving to CCR($\mathcal{H}$), one then arrives at rich families of (mutually inequivalent) CCR-representations. In each representation π, the CCR is then realized in a new Hilbert space, depending on π.

A systematic study of the representations may be undertaken via positive linear functionals (states). This is the so called *Gelfand–Naimark–Segal* (GNS) construction. In the case of CCR($\mathcal{H}$), there is special representation, called the Fock representation (details below), and it is realized in the *symmetric Fock space* over $\mathcal{H}$ (see Section 3.2 for details.).

2.1 Some history and comments on commutation relations

As noted, these commutation relations fall in two kinds, the CCRs and the CARs; the first referring to commutators; and the second to anticommutators, hence the "A" in the CARs. And the algebraic systems based on these respective relations are called the CCR-algebras, and the CAR-algebras. We shall concentrate below on the CCR theory.

CCR. When we speak of the *canonical commutation relations* there is a variety of instances of occurrence, both in pure and applied mathematics. In quantum physics (bosons, or rather Bose–Einstein statistics), the fundamental relation between canonical conjugate quantities (quantities which are related by rules inherent to the statistics of quantum particles or quantum fields. This may refer to position and momentum are vectors of operators, as well as their commutation relations resulting from different components of position and momentum. The CCR's role may be attributed to many pioneers. Here we stress Max Born (1925), who used the term "quantum condition". A number of authors, e.g., E. Kennard (1927) observed that the CCRs imply the Heisenberg uncertainty principle.

Connections to representation theory: For a *finite number of degrees of freedom*, the Stone–von Neumann theorem (the 1950's) asserts uniqueness (up to unitary equivalence) of their representations. It is important that uniqueness for the representations in an *infinite number of degrees of freedom* does not hold.

In **harmonic analysis**, the CCRs underlie the notion of Fourier duality, and they are entailed by definition of variables in Fourier transform theory (time/frequency), or Fourier series (the case of discrete time); or in signal processing.

CAR. As in the case of the CCRs, the CAR algebras (canonical anticommutation relations) play a fundamental role in quantum physics (quantum statistical mechanics, and quantum field theory), specifically, the quantum mechanical study of fermions, or rather Fermi–Dirac statistics. The CARs are equally important in the theory of operator algebras, and in representation theory. Closely related are the algebras of Weyl and Clifford. Weyl and Clifford algebras allow basis-free formulations of the canonical commutation and anticommutation relations, as they relate to symplectic, and a symmetric non-degenerate bilinear forms. In addition, the binary elements in graded Weyl algebras yield a basis-free formulation of the commutation relations of the symplectic and indefinite orthogonal Lie algebras.

This paper [SS65] by Shale and Stinespring covers holomorphic spinor representation of the complex Clifford algebras (closely related to the CAR algebras). With the use of the GNS correspondence between states and representations, Shale [SS64] gives a criterion for equivalence of a pair of representations of a fixed Clifford algebra.

While there are important connections between *commutation relations* and *stochastic analysis* for both classes, CCRs and CARs, we have concentrated here on the case of the CCRs. One reason for this is that the CCRs, yield corresponding random processes which are Gaussian processes, or gaussian fields. By contrast, the random processes associated to the CARs are *determinantal processes*. Unfortunately, details here are beyond the scope of the present book. Nonetheless, we include below a super-condensed summary of some highpoints from the theory of *determinantal measures*; at least as they relate to our main themes. There are many interesting and current papers on *determinantal processes*; see, e.g., [BT14, PV05, Buf16, BQ15, PP14, Lyo03, GP17].

Take as a starting point a fixed positive definite (p.d.) kernel K on some set, say S.

Problem: We want a probability measure $\mathbb{P}$ (called a determinantal probability measure) defined on all subsets Ω of S. This will allow us to study the distribution of subsets of S, also referred to as point distributions; for example the distribution of zeros of Gaussian analytic functions.

Starting with Ω, then the *cylinder subsets* of Ω will be indexed by finite subset F in S as $C(F) := \{X : X \supset F\}$. For every finite subset F in S, we must make sense of determinantal probability measures $\mathbb{P}$, applied to $C(F)$. We need to know that the following assignment $\mathbb{P}(\{X : X \supset F\}) = \det(K|_{F \times F})$ does in fact define a probability measure. So the question is: When does this ansatz define a probability measure $\mathbb{P}$ on cylinder sets $C(F)$? If so, then, of course, it will extended to the cylinder-sigma algebra. In the affirmative case, we can obtain $\mathbb{P}$ via a *Kolmogorov* type *inductive limit construction*. Of course, these cylinder sets define a sigma algebra of subsets of Ω.

Answer: Yes, if the pair (K, μ) is such that K defines a contractive integral operator in $L^2(\mu)$.

The proof is long and beyond our present scope; see the cited references.

Background. Start with a fixed kernel K, assumed p.d. on $S \times S$. There is an interesting theory of *determinantal probability measures* for both the case when S is countable discrete, and also the general case when S is equipped with a sigma algebra, and so forming a measure space. In the first case, take μ to be counting measure, in the second, the relevant choices for μ are only assumed sigma-finite.

To summarize, the p.d. kernels K which admit determinantal probability measures are the ones which allow pairs (K, μ) such that K defines a *contractive integral operator* T_K in $L^2(\mu)$. Most authors also assume that T_K is locally trace class.

So in our language from Chapter 4 below, μ must be generalized Carleson, and so in $GC(K)$ with bound 1. We will obtain a determinantal probability measure for every choice of such μ on S.

2.2 Infinite-dimensional analysis

Our purpose is to identify a unifying framework in infinite-dimensional analysis which involves a core duality notion. There are two elements to our point of view: (i) presentation of the general setting of duality and representation theory, and (ii) detailed applications to two areas, often considered disparate. The first is stochastic analysis, and the second is from the theory of von Neumann algebras (Tomita–Takesaki theory), see, e.g., [BR79] and [BR81]. We feel that this viewpoint is useful as it adds unity to the study of infinite-dimensional analysis; and further, because researchers in one of these two areas usually do not explore the other, or are even unfamiliar with the connections.

We study densely defined unbounded operators acting between different Hilbert spaces, and for these, we introduce a notion of symmetric (closable) pairs of operators. We give applications to themes at the cross road of commutation relations (operator theory) and stochastic calculus. While both subjects have been studied extensively, our aim is to show that the notion of closable pairs from the theory of unbounded operators serves to unify the two areas. Both areas are important in mathematical physics, but researchers familiar with operator theory typically do not appreciate the implications of results on unbounded operators and their commutators for stochastic analysis; and vice versa.

Both the study of quantum fields, and of quantum statistical mechanics, entails families of representations of the canonical commutation relations (CCRs). (For details, see Chapters 3, 4 and 6.) In the case of an infinite number of degrees of freedom, it is known that we have existence of many inequivalent representations of the CCRs. Among the representations, some describe such things as a nonrelativistic infinite free Bose gas of uniform density. But the representations of the CCRs play an equally important role in the kind of infinite-dimensional analysis currently used in a calculus of variation approach to Gaussian fields, Itô integrals, including the Malliavin calculus. In the literature, the infinite-dimensional stochastic operators of derivatives and stochastic integrals are usually taken as the starting point, and the representations of the CCRs are an afterthought. Here we turn the tables. As a consequence of this, we are able to obtain a number of explicit results in an associated multivariable spectral theory. Some of the issues involved are subtle because the operators in the representations under consideration are unbounded (by necessity), and, as a result, one must deal with delicate issues of domains of families of operators and their extensions.

The representations we study result from the Gelfand–Naimark–Segal construction (GNS) applied to certain states on the CCR-algebra. Our conclusions and main results regarding this family of CCR representations (details below, especially Sections 6.3 and 6.4) hold in the general setting of Gaussian fields. But for the benefit of readers, we have also included an illustration dealing with the simplest case, that of the standard Brownian/Wiener process. Many arguments in the special case carry over to general Gaussian fields *mutatis mutandis*. In the Brownian case, our initial Hilbert space will be $\mathscr{L} = L^2(0, \infty)$.

From the initial Hilbert space $\mathscr{L}$, we build the $*$-algebra CCR $(\mathscr{L})$ as in Section 3.1. We will show that the Fock state on CCR $(\mathscr{L})$ corresponds to the Wiener measure $\mathbb{P}$. Moreover the corresponding representation π of CCR $(\mathscr{L})$ will be acting on the Hilbert space $L^2(\Omega, \mathbb{P})$ in such a way that for every k in $\mathscr{L}$, the operator $\pi(a(k))$ is the Malliavin derivative in the direction of k. We caution that the representations of the $*$-algebra CCR $(\mathscr{L})$ are by unbounded operators, but the operators in the range of the representations will be defined on a single common dense domain.

Example: There are two ways to think of systems of generators for the CCR-algebra over a fixed infinite-dimensional Hilbert space ("CCR" is short for canonical commutation relations.):

(i) an infinite-dimensional Lie algebra, or
(ii) an associative $*$-algebra.

With this in mind, (ii) will simply be the universal enveloping algebra of (i); see [Dix77]. While there is also an infinite-dimensional "Lie" group corresponding to (i), so far, we have not found it as useful as the Lie algebra itself.

All this, and related ideas, supply us with tools for an infinite-dimensional stochastic calculus. It fits in with what is called Malliavin calculus, but our present approach is different, and more natural from our point of view; and as corollaries, we obtain new and explicit results in multivariable spectral theory which we feel are of independent interest.

There is one particular representation of the CCR version of (i) and (ii) which is especially useful for stochastic calculus. We call this representation the Fock vacuum-state representation. One way of realizing the representations is abstract: Begin with the Fock vacuum state (or any other state), and then pass to the corresponding GNS representation. The other way is to realize the representation with the use of a choice of a Wiener L^2-space. We prove that these two realizations are unitarily equivalent.

By stochastic calculus we mean stochastic derivatives (e.g., Malliavin derivatives), and integrals (e.g., Itô-integrals). We begin with the task of

realizing a certain stochastic derivative operator as a closable operator acting between two Hilbert spaces.

There is an extensive literature on quantum stochastic calculus based
on the Fock, and other representations, of the CCR including its relation
to Malliavin calculus. The list of authors includes R. Hudson, K. R.
Parthasarathy and collaborators. We refer the reader to the papers [HP84,
AH84, HPP00, CH13, Hud14], and also see [Bog98, HKPS13]. Of more
recent papers dealing with results which have motivated our presentation are [Che13, PS11, LS08, Fre14, LG14, AKL16, AØ15, ZRK15, BZ12,
AJLM15, HS98, Hid07, Hid90, Hid85, Hid71, Hid93].

For GNS constructions, Powers' paper [Pow74] offers a useful identification of classes of representations. With the use of the GNS correspondence
between states and representations, Shale [Sha62] gives a criterion for equivalence of a pair of representations of a fixed CCR algebra, infinite degrees
of freedom.

2.3   Positivity and representations

We begin with a review of some key notions from representation theory,
concluding, in Corollary 2.7, with a key formula for the CCRs. This will
all be needed later in our study of infinite-dimensional analysis, especially
the Malliavin derivative.

Overview. Our focus will be the following:

(i) $*$-algebras $\mathfrak{A}$, states ω on $\mathfrak{A}$ ($\omega\left(A^{*}A\right) \geq 0$, for all $a \in \mathfrak{A}$) and the
corresponding GNS representation on $\mathscr{H}\left(\pi\right)$, with $\pi\left(A\right)\left[B\right] = \left[AB\right]$,
$A, B \in \mathfrak{A}$, and $\left[B\right], \left[AB\right] \in \mathscr{H}\left(\pi\right)$, satisfying

$$\pi\left(A_1 A_2\right) = \pi\left(A_1\right)\pi\left(A_2\right),$$
$$\pi\left(A^{*}\right) = \pi\left(A\right)^{*}.$$

(ii) Lie algebra $\mathfrak{g}$ ($\left[x, y\right] \in \mathfrak{g}$, for all $x, y \in \mathfrak{g}$), representations $\rho \in$
$Rep\left(\mathfrak{g}, V\right)$, V a vector space, with

$$\rho\left(\left[x, y\right]\right) = \left[\rho\left(x\right), \rho\left(y\right)\right]$$
$$= \rho\left(x\right)\rho\left(y\right) - \rho\left(y\right)\rho\left(x\right).$$

(iii) Groups G, p.d. kernels K satisfying $K = K^{g}$ (see (2.1)), and unitary
representations π of G:

$$\pi\left(gg'\right) = \pi\left(g\right)\pi\left(g'\right),$$
$$\pi\left(g\right)^{*} = \pi\left(g^{-1}\right).$$

Representation theory: The case of groups. Definition 2.1. Let $K :$ $S \times S \to \mathbb{C}$ be a p.d. kernel on a group G. Set

$$K^g(s,t) := K(gs, gt), \quad \forall s,t,g \in G. \tag{2.1}$$

The kernel K is said to be G-invariant if $K = K^g$.

Proposition 2.2 (Invariance of p.d. kernels and unitary representations). *Let K be a p.d. kernel on G, and assume that K is G-invariant. Let $\mathscr{H}(K)$ be the corresponding RKHS. For $F \in \mathscr{H}(K)$, let*

$$(\pi(g)F)(s) = F(g^{-1}s), \; g \in G, \tag{2.2}$$

then π defines a unitary representation of G, acting on the Hilbert space $\mathscr{H}(K)$.

Proof. From (2.1), we get the axioms for representations, i.e.,

$$\pi(gg') = \pi(g)\pi(g'), \tag{2.3}$$
$$\pi(g)^* = \pi(g^{-1}), \tag{2.4}$$

for all $g, g' \in G$. In (2.3), the $(\cdot)^*$ refers to the adjoint relative to the $\mathscr{H}(K)$-inner product. $\qquad\square$

Definition 2.3. A unitary representation π of a topological group G is said to be *strongly continuous*, if for all $F \in \mathscr{H}(\pi)$, we have

$$\lim_{g \to e} \|\pi(g)F - F\|_{\mathscr{H}(\pi)} = 0. \tag{2.5}$$

Lemma 2.4. *Let K be a p.d. kernel, and G a topological group such that K is G-invariant (see (2.1)). Then the corresponding unitary representation π (see (2.2)) is strongly continuous if and only if*

$$\lim_{g \to e} K(gs, s) = K(s, s), \; \forall s \in S. \tag{2.6}$$

Proof. Since $\mathscr{H}(K)$ is the Hilbert completion of the kernel functions $K_s(\cdot) = K(\cdot, s)$, it is enough to show that

$$\lim_{g \to e} \|\pi(g)K_s - K_s\|_{\mathscr{H}(K)}^2 = 0, \tag{2.7}$$

for all $s \in S$; see (2.5).

But computation of $\|\cdots\|_{\mathscr{H}(K)}^2$ from (2.7) yields

$$\|\cdots\|_{\mathscr{H}(K)}^2 = 2K(s,s) - 2\Re K(gs,s),$$

and so (2.7) is equivalent to (2.6) as asserted. $\qquad\square$

Given a positive definite kernel K, then a natural question arises: Make precise the notion of *factorization* for K. Now we know that K must be a covariance kernel for a centered Gaussian process (see Chapter 4 for details), so this is one possible factorization. What are the others?

A general discussion begins with Corollary 2.5 below. See also Corollary 4.78 which gives a more complete discussion. Such factorizations also encompasses boundary limits for kernels; see, e.g., Section 8.2.

Our use of the term *factoriz*ation for K in (2.8) is motivated by the notion of factorizations for positive definite matrices.

Corollary 2.5. *Let $K : S \times S \to \mathbb{C}$ be a p.d. kernel with a group G satisfying the invariance condition* (2.1). *Suppose $(X, \mathscr{B}, \mu, \psi)$ are such that*

$$K(s,t) = \int_X \overline{\psi(s,x)} \psi(t,x) \, \mu(dx), \qquad (2.8)$$

i.e., K has a measurable factorization.

Then the representation of G action in $\mathscr{H}(K)$, see (2.2), *induces a representation $\widetilde{\pi}$ acting instead in the subspace $L(\mathscr{H}(K)) \subseteq L^2(\mu)$, given by*

$$\widetilde{\pi}(g)\psi(s,\cdot) = \psi(g^{-1}s,\cdot),$$

where $s \in S$, $g \in G$. Here, L denotes the isometry $\mathscr{H}(K) \to L^2(\mu)$ in (2.9).

Proof. The conclusion follows from the canonical property of the isometry $L : \mathscr{H}(K) \to L^2(\mu)$, specified by

$$L(K(s, \underset{\substack{\downarrow \\ \text{in } S}}{\cdot}\,))(x) = \psi(s, \underset{\substack{\downarrow \\ \text{in } X}}{x}) \qquad (2.9)$$

and extended by linearity and $\|\cdot\|_{\mathscr{H}(K)}$ norm closure. Note that L will be isometric when so extended, which follows from (2.8).

We are assuming that K is G-invariant. So we get two versions of the representation π of G, one acting on the Hilbert space $\mathscr{H}(K)$, and the other on $L(\mathscr{H}(K)) \subset L^2(\mu)$; and L yields an intertwining operator for the two representations (Figure 2.1). $\qquad\square$

Three classes of factorizations. Given a p.d. kernel K on a suitable set, say S, then the corresponding class of factorizations, as in (2.8) in Corollary 2.5, will play an important role in subsequent chapters. It turns out that there is a rich variety of choices, measure space X, and associated data, $(X, \mathscr{B}, \mu, \psi)$. Some of the important choices are as follows: We list

$$\begin{array}{ccc} \mathscr{H}(K) & \xrightarrow{\quad L \quad} & L(\mathscr{H}(K)) \subset L^2(\mu) \\ {\scriptstyle \widetilde{\pi}(g)} \big\downarrow & & \big\downarrow {\scriptstyle \widetilde{\pi}(g)} \\ \mathscr{H}(K) & \xrightarrow[\quad L \quad]{} & L(\mathscr{H}(K)) \subset L^2(\mu) \end{array}$$

Figure 2.1: L as an intertwining operator

three, and they will be discussed in detail in subsequent chapters: (i) In the setting of classical harmonic analysis, and in metric geometry, it is of interest to identify factorizations (2.8) where X may be chosen to be a suitable metric boundary of S. (ii) In applications to stochastic processes, one studies probabilistic boundaries. It is then natural to identify Gaussian processes, indexed by S, and having K as covariance kernel. In this case, of course X will chosen to be the corresponding probability space, Ω, equipped with the cylinder sigma-algebra, and probability measure $\mathbb{P}$ which realize this covariance calculation for K. (iii) For an analysis of ONBs or Parseval frames in $\mathscr{H}(K)$, the RKHS, one may take for X the natural numbers $\mathbb{N}$ with counting measure.

2.4 An infinite-dimensional Lie algebra

Let $\mathscr{L}$ be a separable Hilbert space, i.e., $\dim \mathscr{L} = \aleph_0$, and let CCR$(\mathscr{L})$ be the corresponding CCR-algebra. As above, its generators are denoted $a(k)$ and $a^*(l)$, for $k, l \in \mathscr{L}$. We shall need the following:

Proposition 2.6.

(i) *The "quadratic" elements in CCR$(\mathscr{L})$ of the form $a(k)\, a^*(l)$, $k, l \in \mathscr{L}$, span a Lie algebra $\mathfrak{g}(\mathscr{L})$ under the commutator bracket.*

(ii) *We have*

$$\begin{aligned} &[a(h)a^*(k), a(l)a^*(m)] \\ &\quad = \langle h, m \rangle_{\mathscr{L}}\, a(l)a^*(k) - \langle l, k \rangle_{\mathscr{L}}\, a(h)a^*(m), \end{aligned}$$

for all $h, k, l, m \in \mathscr{L}$.

(iii) *If $\{\varepsilon_i\}_{i \in \mathbb{N}}$ is an ONB in $\mathscr{L}$, then the non-zero commutators are as follows: Set $\gamma_{i,j} := a(\varepsilon_i)a^*(\varepsilon_j)$, then, for $i \neq j$, we have*

$$[\gamma_{i,i}, \gamma_{j,i}] = \gamma_{j,i}; \tag{2.10}$$

$$[\gamma_{i,i}, \gamma_{i,j}] = -\gamma_{i,j}; \quad \textit{and} \tag{2.11}$$

$$[\gamma_{j,i}, \gamma_{i,j}] = \gamma_{i,i} - \gamma_{j,j}. \tag{2.12}$$

All other commutators vanish; in particular, $\{\gamma_{i,i} \mid i \in \mathbb{N}\}$ spans an abelian sub-Lie algebra in $\mathfrak{g}(\mathscr{L})$.

Note further that, when $i \neq j$, then the three elements

$$\gamma_{i,i} - \gamma_{j,j}, \quad \gamma_{i,j}, \quad and \quad \gamma_{j,i} \tag{2.13}$$

span (over $\mathbb{R}$) an isomorphic copy of the Lie algebra $sl_2(\mathbb{R})$.

(iv) *The Lie algebra generated by the first-order elements $a(h)$ and $a^*(k)$ for $h, k \in \mathscr{L}$, is called the Heisenberg Lie algebra $\mathfrak{h}(\mathscr{L})$. It is normalized by $\mathfrak{g}(\mathscr{L})$; indeed we have:*

$$[a(l)a^*(m), a(h)] = -\langle m, h\rangle_{\mathscr{L}}\, a(l), \quad and$$
$$[a(l)a^*(m), a^*(k)] = \langle l, k\rangle_{\mathscr{L}}\, a^*(m), \quad \forall l, m, h, k \in \mathscr{L}.$$

Proof. The verification of each of the four assertions (i)–(iv) uses only the fixed axioms for the CCR, i.e.,

$$\begin{cases} [a(k), a(l)] & = 0, \\ [a^*(k), a^*(l)] = 0, \quad and \\ [a(k), a^*(l)] & = \langle k, l\rangle_{\mathscr{L}}\, \mathbb{1}, \quad k, l \in \mathscr{L}; \end{cases} \tag{2.14}$$

where $\mathbb{1}$ denotes the unit-element in $CCR(\mathscr{L})$. $\quad\square$

Corollary 2.7. *Let $CCR(\mathscr{L})$ be the CCR-algebra with generators $a(k)$, $a^*(l)$, $k, l \in \mathscr{L}$, and let $[\cdot, \cdot]$ denote the commutator Lie bracket; then, for all $k, h_1, \ldots, h_n \in \mathscr{L}$, and all $p \in \mathbb{R}[x_1, \ldots, x_n]$ (the n-variable polynomials over $\mathbb{R}$), we have*

$$[a(k), p(a^*(h_1), \ldots, a^*(h_n))]$$
$$= \sum_{i=1}^{n} \frac{\partial p}{\partial x_i}(a^*(h_1), \ldots, a^*(h_n))\langle k, h_i\rangle_{\mathscr{L}}. \tag{2.15}$$

Proof. The verification of (2.15) uses only the axioms for the CCR, i.e., the commutation relations (2.14) above, plus a little combinatorics. $\quad\square$

We shall return to a stochastic variation of formula (2.15), the so called Malliavin derivative in the direction k. In this, the system $(a^*(h_1), \ldots, a^*(h_n))$ in (2.15) instead takes the form of a multivariate Gaussian random variable.

2.5 Guide to the literature

While we have cited some paper from the Bibliography in our discussion inside the chapter, readers not familiar with the ideas involved may wish to consult the following sources: [AW73, Arv76a, AW63, Che13, Dix77, GJ87, Hei69, Hid80, Pol02, Sak98].

Chapter 3

Representation Theory, with Emphasis on the Case of the CCRs

> "*Probability is a mathematical discipline whose aims are akin to those, for example, of geometry of analytical mechanics. In each field we must carefully distinguish three aspects of the theory: (a) the formal logical content, (b) the intuitive background, and (c) the applications. The character, and the charm, of the whole structure cannot be appreciated without considering all three aspects in their proper relation.*" Feller, William; An Introduction to Probability Theory and its Applications.

While the subject of representation theory is vast, with many and varied subdisciplines, and with many and diverse applications, for our present purposes we have made choices as follows, each motivated by the aim of the book: the topics include transform theory in the analysis on Reproducing Kernel Hilbert Space (RKHS) (Chapter 4), numerical models (Section 4.3), Gaussian Hilbert space (Section 4.4), Itô integrals (Section 3.3), selected classes of stochastic differential equations (SDEs) (see, e.g., a discussion in Section 3.2.) Useful supplement treatments in the literature: [AH84, BR81, Pow74].

For given algebraic system, say a group, or a non-commutative algebra (such as a Lie algebra), there is an associated theory of representations. By representation we mean a realization by operators in Hilbert space; more precisely a homomorphism from the prescribed algebraic system into a corresponding system (group, algebra etc.) of operators in Hilbert space.

It is only when we pass to representations that Hilbert space and operators enter the discussion. We stress that infinite dimensions is forced on us by applications. Infinite dimensions are "fun," but also necessary.

3.1 The CCR-algebra, and the Fock representations

There are two $*$-algebras built functorially from a fixed (single) Hilbert space $\mathscr{L}$; often called the one-particle Hilbert space (in physics). The dimension $\dim \mathscr{L}$ is called *the number of degrees of freedom*. The case of interest here is when $\dim \mathscr{L} = \aleph_0$ (countably infinite). The two $*$-algebras are called the CAR, and the CCR-algebras, and they are extensively studied; see, e.g., [BR81]. Of the two, only $\mathrm{CAR}(\mathscr{L})$ is a C^*-algebra. The operators arising from representations of $\mathrm{CCR}(\mathscr{L})$ will be *unbounded*, but still having a common dense domain in the respective representation Hilbert spaces. In both cases, we have a Fock representation. For $\mathrm{CCR}(\mathscr{L})$, it is realized in the symmetric Fock space $\Gamma_{sym}(\mathscr{L})$. There are many other representations, inequivalent to the respective Fock representations.

It is useful to make a distinction between (i) the CCR as an algebra with involution; and (ii) the representations of the CCR.

In both cases, the starting point is a fixed Hilbert space, say $\mathscr{L}$. From this, we build a $*$-algebra of symbols $a(h)$ and $a^*(k)$, where $h, k \in \mathscr{L}$. The $\mathrm{CCR}(\mathscr{L})$ is generated axiomatically by a system, $a(h)$, $a^*(h)$, $h \in \mathscr{L}$, subject to

$$[a(h), a(k)] = 0, \quad \forall h, k \in \mathscr{L}, \text{ and}$$

$$[a(h), a^*(k)] = \langle h, k \rangle_{\mathscr{L}} \, \mathbb{1}. \tag{3.1}$$

Here, $\mathbb{1}$ is the unit in $\mathfrak{A}(\mathscr{L}) = \mathrm{CCR}(\mathscr{L})$; and where the inner product is taken to be linear in the second variable. The involution $A \to A^*$ in $\mathfrak{A}(\mathscr{L})$ is denoted by $(a(h))^* = a^*(h)$, for $h \in \mathscr{L}$.

Notation. In (3.1), $[\cdot, \cdot]$ denotes the commutator. More specifically, if A, B are elements in a $*$-algebra, set $[A, B] := AB - BA$.

To specify a representation of $\mathfrak{A}(\mathscr{L})$, we must have a Hilbert space, say $\mathscr{F}$, and a homomorphism $\pi : \mathfrak{A}(\mathscr{L}) \to$ operators on $\mathscr{F}$. For $A \in \mathfrak{A}(\mathscr{L})$, we require that

$$\pi(A^*) = \pi(A)^*, \tag{3.2}$$

where the $*$ on the right-hand side of (3.2) refers to the adjoint operator.

It is a fact that the CCR algebra $\mathfrak{A}(\mathscr{L})$ does not have representations by bounded operators, see, e.g., [BR81]. Nonetheless, when $(\pi, \mathscr{F})$ is a representation, there is a natural choice of dense subspace $\mathscr{D}(\pi) \subset \mathscr{F}(\pi)$, such that all the operators $\{\pi(A) : A \in \mathfrak{A}(\mathscr{L})\}$ are defined on $\mathscr{D}(\pi)$. For the case of the Fock representation π_{Fock} considered below, the choice of subspace $\mathscr{D}(\pi_{Fock})$ will be clear.

But in the case of a general representation, one makes use of positive linear functionals ω on $\mathfrak{A}(\mathscr{L})$, and the associated Gelfand–Naimark–Segal (GNS) representation; again, see [BR81] or [GJ87] for details. In rough outline: A linear functional $\omega : \mathfrak{A}(\mathscr{L}) \to \mathbb{C}$ is said to be positive if

$$\omega(A^*A) \geq 0 \tag{3.3}$$

for all $A \in \mathfrak{A}(\mathscr{L})$. If further $\omega(\mathbb{1}) = 1$, we say that ω is a *state*.

The GNS representation π_ω. Start with a state ω, the corresponding representation $\pi = \pi_\omega$ is specified by a Hilbert space $\mathscr{F}_\omega$ and a (cyclic) vector $v \in \mathscr{F}_\omega$, such that

$$\omega(A) = \langle v, \pi(A)v \rangle_{\mathscr{F}}, \quad A \in \mathfrak{A}(\mathscr{L}). \tag{3.4}$$

The construction of $(\mathscr{F}_\omega, \pi_\omega)$ for a given state ω is similar to the construction of a RKHS from a given positive definite kernel K. Start with the following pre-inner product

$$\mathfrak{A}(\mathscr{L}) \times \mathfrak{A}(\mathscr{L}) : (A, B) \longmapsto \omega(A^*B), \tag{3.5}$$

using (3.3), we get the Schwarz inequality

$$|\omega(A^*B)|^2 \leq \omega(A^*A)\,\omega(B^*B). \tag{3.6}$$

Define

$$\pi(A)B = AB, \tag{3.7}$$

as a product in $\mathfrak{A}(\mathscr{L})$, and then extend to the Hilbert completion arising from (3.5). For cyclic vector $v\,(= v_\omega)$, we take $v = 1$ but viewed as a vector in $\mathscr{F}$ (the Hilbert completion).

Combining (3.6) and (3.7), we see that $\{\pi(A)\}_{A \in \mathfrak{A}(\mathscr{L})}$ turns into operators, and that (3.4) will then hold.

For dense subspace $\mathscr{D}(\pi)$ in $\mathscr{F}(\pi)$, we take

$$\mathscr{D}(\pi) = \{\pi(A)v\}_{A \in \mathfrak{A}(\mathscr{L})}.$$

The special case of the Fock representation π_{Fock} arise from the case when the state is taken to be $\omega = \omega_{Fock}$; see (3.8) & (3.9). The Hilbert space $\mathscr{F}(\pi_{Fock})$ is the symmetric Fock space over $\mathscr{L}$; see details below.

The *Fock States* ω_{Fock} on the CCR-algebra are specified as follows:

$$\omega_{Fock}(a\,(h)\,a^*\,(k)) = \langle h, k \rangle_{\mathscr{L}} \qquad (3.8)$$

with the vacuum property

$$\omega_{Fock}(a^*(h)a(h)) = 0, \ \forall h \in \mathscr{L}. \qquad (3.9)$$

For the corresponding Fock representations π we have:

$$[\pi(h), \pi^*(k)] = \langle h, k \rangle_{\mathscr{L}} \, I_{\Gamma_{sym}(\mathscr{L})}, \qquad (3.10)$$

where $I_{\Gamma_{sym}(\mathscr{L})}$ on the RHS of (3.10) refers to the identity operator.

Some relevant papers regarding the CCR-algebra and its representations are [AW63, Arv76a, Arv76b, PS72a, PS72b, AW73, GJ87, JP91].

3.2   Symmetric Fock space and Malliavin derivative

The *Malliavin derivative* is the key ingredient in what goes by the name, Malliavin calculus. It is a mathematical framework for infinite-dimensional analysis. It is motivated in part by calculus of variations, and optimization problems, from the classical case of deterministic functions, but now extended to calculus for stochastic processes. It begins with a precise formulation of derivatives of random variables. Malliavin calculus is also called the stochastic calculus of variations.

We begin the discussion with an outline of a few facts about the symmetric Fock space. For additional and related material see Lemma 3.4 below, and Section 2.2 above.

The Fock space is an orthogonal expansion of tensor Hilbert spaces; a construction popular in quantum mechanics for modelling quantum states. It is named after V. A. Fock[1] who introduced it in his 1932 paper "Konfigurationsraum und zweite Quantelung". Technically, the Fock space is (the Hilbert space completion of) the direct sum of the symmetric or antisymmetric tensors in the tensor powers of a single-particle Hilbert space $\mathscr{H}$. In more detail, a Fock space is the sum of a set of Hilbert spaces representing zero particle states, one particle states, two particle states, and so on. If the identical particles are bosons (governed by the CCRs), the n-particle states are vectors in a symmetrized tensor product of n single-particle Hilbert spaces $\mathscr{H}$. If the identical particles are fermions (governed by the CARs), then the n-particle states are vectors in an antisymmetrized tensor product of n single-particle Hilbert spaces $\mathscr{H}$.

---

[1]See the short bio in Appendix A.

Let $\mathscr{L}$ and $\mathscr{H}$ be two Hilbert spaces. Let $\Gamma_{sym}(\mathscr{H})$ and $\Gamma_{sym}(\mathscr{L})$ be the corresponding symmetric Fock spaces. Set

$$\varepsilon(h) = e^h = \sum_{n=0}^{\infty} \frac{1}{\sqrt{n!}} \underbrace{h \otimes \cdots \otimes h}_{n \text{ times}} = \sum_{n=0}^{\infty} \frac{h^{\otimes n}}{\sqrt{n!}}. \tag{3.11}$$

Lemma 3.1. *For $h_1, h_2 \in \mathscr{H}$, we have*

$$\langle e^{h_1}, e^{h_2} \rangle_{\Gamma_{sym}(\mathscr{H})} = \sum_{n=0}^{\infty} \frac{\langle h_1, h_2 \rangle_{\mathscr{H}}^n}{n!} = \exp(\langle h_1, h_2 \rangle_{\mathscr{H}}). \tag{3.12}$$

In particular,

$$\left\| e^h \right\|_{\Gamma_{sym}(\mathscr{H})}^2 = \exp\left(\|h\|_{\mathscr{H}}^2 \right), \quad \text{for all } h \in \mathscr{H}. \tag{3.13}$$

Proof. Formula (3.12) follows from the basic definition of $\Gamma_{sym}(\mathscr{H})$, where $\langle f^{\otimes n}, h^{\otimes m} \rangle_{\Gamma_{sym}(\mathscr{H})} = 0$ if $n \neq m$, and $\langle f^{\otimes n}, h^{\otimes n} \rangle_{\Gamma_{sym}(\mathscr{H})} = \langle f, h \rangle_{\mathscr{H}}^n$. Hence,

$$\langle \varepsilon(f), \varepsilon(h) \rangle_{\Gamma_{sym}(\mathscr{H})} = \langle e^f, e^h \rangle_{\Gamma_{sym}(\mathscr{H})}$$

$$= \left\langle \sum_{n=0}^{\infty} \frac{f^{\otimes n}}{\sqrt{n!}}, \sum_{m=0}^{\infty} \frac{h^{\otimes m}}{\sqrt{m!}} \right\rangle_{\Gamma_{sym}(\mathscr{H})}$$

$$= \sum_{n=0}^{\infty} \frac{\langle f^{\otimes n}, h^{\otimes n} \rangle_{\Gamma_{sym}(\mathscr{H})}}{n!}$$

$$= \sum_{n=0}^{\infty} \frac{\langle f, h \rangle_{\mathscr{H}}^n}{n!} = \exp(\langle f, h \rangle_{\mathscr{H}}). \qquad \square$$

Lemma 3.2. *The $\Gamma_{sym}(\mathscr{H})$ norm agrees with the RKHS norm for the p.d. function*

$$\Gamma_{sym}(f, g) := \exp(\langle f, g \rangle_{\mathscr{H}}), \quad f, g \in \mathscr{H}.$$

Proof. Let $\{c_i\}$, $\{f_i\}$, $c_i \in \mathbb{R}$ (or $\mathbb{C}$), $f_i \in \mathscr{H}$ be given; then

$$\left\| \sum_i c_i e^{f_i} \right\|_{\Gamma_{sym}(\mathscr{H})}^2 = \sum_i \sum_j c_i c_j \langle e^{f_i}, e^{f_j} \rangle_{\Gamma_{sym}(\mathscr{H})}$$

$$= \sum_i \sum_j c_i c_j \exp\left(\langle f_i, f_j \rangle_{\mathscr{H}} \right),$$

which is the RKHS norm as stated. $\qquad \square$

As discussed above, we may view CCR as abstract algebras (A) and as representations (Rep). Let $\mathfrak{A} = \mathrm{CCR}(\mathcal{H})$ be the CCR algebra.

Definition 3.3. We say that ω is a vacuum state if

$$\omega(a^*(h)a(h)) = 0, \quad \forall h \in \mathcal{H}. \tag{3.14}$$

When we pass to the GNS representation π with given vector $v_0 \in \mathcal{H}(\pi)$, then

$$\omega(A) = \langle v_0, \pi(A)v_0 \rangle_{\mathcal{H}(\pi)}, \quad A \in \mathfrak{A};$$

so

$$
\begin{aligned}
\omega\left(a^*\left(h\right) a\left(h\right)\right) &= \langle v_0, \pi\left(a^*\left(h\right) a\left(h\right)\right) v_0 \rangle_{\mathcal{H}(\pi)} \\
&= \langle \pi\left(a\left(h\right)\right) v_0, \pi\left(a\left(h\right)\right) v_0 \rangle_{\mathcal{H}(\pi)} \\
&= \|\pi\left(a\left(h\right)\right) v_0\|^2_{\mathcal{H}(\pi)} = 0,
\end{aligned}
$$

and so $v_0 \in ker\,(\text{annihilation})$.

Note there are many representations, not satisfying (3.14), i.e., not vacuum. In the *Fock representation* we get corresponding the vector $a\left(h\right) v_0$ vanishing precisely $\pi\left(a\left(h\right)\right) v_0 = 0$. The Hilbert space $\mathcal{H}(\pi)$ of the Fock representation is the *symmetric Fock space*. See the table below.

Table 3.1: Functions, from Hilbert space to algebra, and to symmetric Fock space.

$\mathcal{H}(\pi)$	$\Gamma_{sym}(\mathcal{H})$
$\pi\left(a^*\left(h\right)^n\right) v_0$	$h^{\otimes h} = \underbrace{h \otimes \cdots \otimes h}_{n},\ n > 0$
$n = 0,\ v_0$	the $n = 0$ vector
$\sum_{n=0}^{\infty} \frac{1}{\sqrt{n!}} \pi\left(a^*\left(h\right)\right) v_0$	$\varepsilon(h) = e^h = \sum_{n=0}^{\infty} \frac{1}{\sqrt{n!}} h^{\otimes n}$

$\mathcal{H}$ Hilbert space	$\longrightarrow$	$\mathfrak{A}(\mathcal{H})$ CCR-algebra	$\longrightarrow$	$\Gamma_{sym}(\mathcal{H})$ symmetric Fock space

In the Fock representation it is convenient to make the identification

$$h \longleftrightarrow a^*(h)$$

for $h \in \mathscr{H}$, but we may also use

$$\pi_{Fock}(a^*(h)) \longleftrightarrow h.$$

Hence

$$a^*(h)^n \longleftrightarrow a^*(h)^n v_0 \longleftrightarrow h^{\otimes h} = \underbrace{h \otimes \cdots \otimes h}_{n}.$$

For the definition of the Fock representation, we then get

$$\begin{aligned}
\langle \varepsilon(h), \varepsilon(f) \rangle_{\Gamma_{sym}(\mathscr{H})} &= \langle e^h, e^f \rangle_{\Gamma_{sym}(\mathscr{H})} \\
&= \sum_{n=0}^{\infty} \frac{\langle h^{\otimes n}, f^{\otimes n} \rangle_{\Gamma_{sym}(\mathscr{H})}}{n!} \\
&= \sum_{n=0}^{\infty} \frac{\langle h, f \rangle_{\mathscr{H}}^n}{n!} \\
&= \exp(\langle h, f \rangle_{\mathscr{H}})
\end{aligned}$$

for all $f, h \in \mathscr{H}$.

Lemma 3.4. *Fix a real Hilbert space $\mathscr{H}$, and consider $Q(h, k) := \exp(\langle h, k \rangle_{\mathscr{H}})$, for $h, k \in \mathscr{H}$. Let $\mathscr{H}(Q)$ be the corresponding RKHS. Then*

$$\mathscr{H}(Q) = \Gamma_{sym}(\mathscr{H}).$$

Proof. Let $h, k \in \mathscr{H}$, then

$$Q(\cdot, k) = \exp(\langle \cdot, k \rangle_{\mathscr{H}}) = \langle \varepsilon(\cdot), \varepsilon(k) \rangle_{\Gamma_{sym}(\mathscr{H})}.$$

Then we have

$$\begin{aligned}
\langle \varepsilon(h), \varepsilon(k) \rangle_{\Gamma_{sym}(\mathscr{H})} &= \left\langle \sum_{n=0}^{\infty} \frac{h^{\otimes n}}{\sqrt{n!}}, \sum_{n=0}^{\infty} \frac{k^{\otimes n}}{\sqrt{m!}} \right\rangle_{\Gamma_{sym}(\mathscr{H})} \\
&= \sum_{n=0}^{\infty} \frac{\langle h, k \rangle_{\mathscr{H}}^n}{n!} = Q(h, k). \qquad \square
\end{aligned}$$

Lemma 3.5. *We have*

$$\mathbb{E}(e^{W_h}) = e^{\frac{1}{2}\|h\|_{\mathscr{H}}^2}.$$

Proof. We use the fact that $\mathbb{E}(W_h^n) = 0$ if n is odd, and

$$\mathbb{E}(W_h^{2n}) = \frac{(2n)!}{2^n n!}\|h\|_{\mathscr{H}}^2 = (2n-1)!!\|h\|_{\mathscr{H}}^2.$$

By expanding the power series, we have

$$\mathbb{E}\left(e^{W_h}\right) = \sum_{n=0}^{\infty} \frac{1}{(2n)!} \mathbb{E}\left(W_h^{2n}\right)$$

$$= \sum_{n=0}^{\infty} \frac{1}{2^n n!} \|h\|^{2n}$$

$$= \exp\left(\frac{1}{2} \|h\|_{\mathscr{H}}^2\right).$$

See Corollary 3.10 for the special case when $\mathscr{H} = L^2(\mu)$. □

Starting with $\mathscr{H}$, there is also a Gaussian process $\{W_h\}$ indexed by $\mathscr{H}$, where $\mathbb{E}(W_h) = 0$ for all $h \in \mathscr{H}$, and

$$\mathbb{E}(W_h W_f) = \langle h, f \rangle_{\mathscr{H}}. \tag{3.15}$$

If $\mathscr{H}$ is a complex Hilbert space, then W_h may be $\mathbb{C}$-valued and (3.15) should be modified as

$$\mathbb{E}\left(\overline{W_h} W_f\right) = \langle h, f \rangle_{\mathscr{H}},$$

for all $h, f \in \mathscr{H}$.

The Gaussian process $\{W_f\}_{f \in \mathscr{H}}$ exists for every Hilbert space $\mathscr{H}$. In the general case, it is simply associated with the inner product $\langle \cdot, \cdot \rangle_{\mathscr{H}}$, which is p.d. on $\mathscr{H} \times \mathscr{H}$. Then there are standard ways of constructing the probability space $(\Omega, \mathscr{C}, \mathbb{P})$ with expectation $\mathbb{E}(\cdots) = \int_{\Omega}(\cdots)\,d\mathbb{P}$, such that $\mathbb{E}(W_f W_g) = \langle f, g \rangle_{\mathscr{H}}$, for all $f, g \in \mathscr{H}$. (We consider here the real case, and the case of complex Hilbert space is similar.)

A special case is $\mathscr{H} = L^2(S, \mathscr{B}_S, \mu)$ with its Hilbert norm $\|\cdot\|_{L^2(\mu)}$. Then $\{W_A : A \in \mathscr{B}_S\}$ is a Gaussian process and $W_f^{(\mu)} = \int f(x)\,dW_x^{(\mu)}$ is an Itô-integral. See Section 3.3 for details.

Lemma 3.6. *For $W_f \in L^2(\Omega, \mathbb{P})$, set*

$$Z\left(e^f\right) := e^{W_f - \frac{1}{2}\|f\|^2}, \quad f \in \mathscr{H}.$$

Then $Z : \Gamma_{sym}(\mathscr{H}) \to L^2(\Omega, \mathbb{P})$ extends to an isometric isomorphism; also intertwining.

Proof. We must prove that when Z is extended by linearity and closure, then it is isometric and onto. This follows from the standard arguments, and the key step is Lemma 3.7 below. □

Lemma 3.7. *We have*

$$\left\langle Z\left(e^f\right), Z\left(e^h\right)\right\rangle_{L^2} = \left\langle e^f, e^h\right\rangle_{\Gamma_{sym}(\mathcal{H})}. \tag{3.16}$$

Proof. We have

$$
\begin{aligned}
\mathrm{RHS}_{(3.16)} &= \exp\left(\langle f, h\rangle_{\mathcal{H}}\right) \\
\mathrm{LHS}_{(3.16)} &= \left\langle e^{W_f - \frac{1}{2}\|f\|^2}, e^{W_h - \frac{1}{2}\|h\|^2}\right\rangle_{L^2} \\
&= \mathbb{E}\left(e^{W(f+h)}\right) e^{-\frac{\|f\|^2 + \|h\|^2}{2}} \\
&\overset{(\text{Lem. } 3.5)}{=} e^{\frac{1}{2}\|f+h\|^2} e^{-\frac{\|f\|^2 + \|h\|^2}{2}} \\
&= \exp(\langle f, h\rangle).
\end{aligned}
$$

$\square$

Lemma 3.8. *Let $Z : \Gamma_{sym}\left(\mathcal{H}\right) \to L^2\left(\Omega, \mathbb{P}\right)$ be as above, then*

$$Z^*\left(e^{W_f}\right) = e^f e^{\frac{1}{2}\|f\|^2_{\mathcal{H}}}, \quad f \in \mathcal{H}.$$

Proof. Indeed, we have

$$
\begin{aligned}
\left\langle Z^*\left(e^{W_f}\right), e^h\right\rangle_{\Gamma_{sym}(\mathcal{H})} &= \mathbb{E}\left(e^{W_f} e^{W_h}\right) e^{-\frac{1}{2}\|h\|^2} \\
&= e^{\frac{1}{2}\|f+h\|^2} e^{-\frac{\|f\| + \|h\|}{2}} e^{\frac{1}{2}\|f\|} \\
&= e^{\langle f, h\rangle} e^{\frac{1}{2}\|f\|} \\
&= \left\langle e^f, e^h\right\rangle_{\Gamma_{sym}(\mathcal{H})} e^{\frac{1}{2}\|f\|} \\
&= \left\langle e^f e^{\frac{1}{2}\|f\|^2}, e^h\right\rangle_{\Gamma_{sym}(\mathcal{H})}.
\end{aligned}
$$

$\square$

A key significance of Itô's lemma is that it makes precise the distinction between the role of the first terms in the Taylor expansion, thus contrasting the familiar deterministic case with the stochastic counterpart, applicable to Gaussian processes. The latter (the stochastic counterpart) entails the quadratic variation. We outline the conclusion below.

Generalized Itô lemma. We also have an extension of the Itô lemma for the process $W_f^{(\mu)}$. Let $\psi \in C^2$, then

$$d\psi(W_f^{(\mu)}) = \psi'(W_f^{(\mu)})dW^{(\mu)} + \frac{1}{2}\psi''(W_f^{(\mu)})d\mu; \tag{3.17}$$

or stated in integral form:

$$\psi(W_f^{(\mu)}) = \int \psi'(W_f^{(\mu)}) f(\cdot) dW^{(\mu)} + \frac{1}{2} \int \psi''(W_f^{(\mu)}) d\mu. \qquad (3.18)$$

Remark 3.9. Note that all parts of the formulas in (3.17) and (3.18) are random variables on the probability space $(\Omega, \mathbb{P})$.

3.3 Itô-integrals

Overview

$\begin{array}{c}(M, \mathscr{B}, \mu) \\ \mu \ \sigma\text{-finite}\end{array}$	$\rightsquigarrow$	$\begin{array}{c}\text{Wiener process} \\ \left\{W_A^{(\mu)}\right\}_{A \in \mathscr{B}_{fin}}\end{array}$	$\rightsquigarrow$	$\begin{array}{c}\text{Itô-integral} \\ \Phi(f) = \int_M f(x)\, dW_x^{(\mu)}\end{array}$

Given a measure space $(M, \mathscr{B}, \mu)$, where μ is σ-finite, we shall construct an associated Wiener process, $W_A^{(\mu)}$, indexed by

$$\mathscr{B}_{fin} = \{A \in \mathscr{B} : \mu(A) < \infty\},$$

and the corresponding Itô-integral

$$\Phi(f) = \int_M f(x)\, dW_x^{(\mu)}, \quad f \in L^2(\mu), \qquad (3.19)$$

satisfying

$$\mathbb{E}(|\Phi(f)|^2) = \int_M |f(x)|^2 d\mu(x). \qquad (3.20)$$

Equation (3.20) is referred to as the Itô-isometry.

Details. The justification for the Itô-integral in (3.19) subtle, since the Wiener process on the RHS in the expression is not of bounded variation. So the alternative approach to stochastic integration, due to Itô, is to instead make use of an isometry-property, see (3.23), and then conclude that the Itô-integral is well defined as an L^2 Gaussian process, i.e., as a unique limit process in $L^2(\Omega, \mathbb{P})$; see (3.24). Starting with a fixed measure μ, we then get a realization of $L^2(\mu)$ as a Gaussian Hilbert space. More general Gaussian Hilbert spaces will be considered in Section 3.3.3 below.

Step 1. The Wiener process $\{W_A : A \in \mathscr{B}_{fin}\}$ is Gaussian, with covarance

$$\mathbb{E}(W_A^{(\mu)} W_B^{(\mu)}) = \mu(A \cap B), \quad \forall A, B \in \mathscr{B}_{fin}, \qquad (3.21)$$

and centered, $\mathbb{E}(W_A^{(\mu)}) = 0$.

Step 2. Every $f \in L^2(\mu)$ admits an approximation over a filter (A_i) of partitions, i.e., $A_i \in \mathscr{B}_{fin}$, $x_i \in A_i$, $A_i \cap A_j = \emptyset$ for $i \neq j$, such that

$$\lim \left\| f - \sum_i f(x_i)\chi_{A_i}(\cdot) \right\|_{L^2(\mu)} = 0, \tag{3.22}$$

where the limit is taken over filters of partitions, and χ_A denotes the indicator function of a specified set A.

Step 3. Fix a system $\{A_i\}$, $x_i \in A$, as specified in (3.22); then

$$\mathbb{E}\left(\left|\sum_i f(x_i)W_{A_i}^{(\mu)}\right|^2\right) = \sum_i |f(x_i)|^2 \mu(A_i); \tag{3.23}$$

since $\mathbb{E}(W_{A_i}^{(\mu)} W_{A_j}^{(\mu)}) = \delta_{ij}\mu(A_i)$.

Step 4. Let $(\Omega, \mathscr{C}, \mathbb{P})$ be the probability space for (3.21), i.e., $\mathbb{E}_\mu(\cdot\cdot) = \int_\Omega (\cdot\cdot)\, d\mathbb{P}$ is the expectation used in (3.21). It follows from (3.22) and (3.23), that the system

$$\left\{\sum_i f(x_i)W_{A_i}^{(\mu)}\right\} \subset L^2(\Omega, \mathbb{P})$$

has a limit in $L^2(\Omega, \mathbb{P})$ over the filter of all partitions. Moreover, this limit is the Itô-integral

$$\Phi(f) := \lim \sum_i f(x_i)\, W_{A_i}^{(\mu)} \in L^2(\Omega, \mathbb{P}), \tag{3.24}$$

and we shall write $\Phi(f) = \int_M f(x)dW_x^{(\mu)}$. It is also immediate from (3.22) and (3.23) that (3.20) holds.

Corollary 3.10. *The formula*

$$\mathbb{E}(e^{\Phi(f)}) = e^{\frac{1}{2}\|f\|^2}, \quad f \in \mathscr{H} = L^2(\mu)$$

follows from the usual approach to the Itô-integral.

Proof. By (3.24), we have

$$\prod_i \mathbb{E}\left(e^{c_i W_{A_i}}\right) = \prod c_i^2 \frac{1}{(2n)!}\mathbb{E}\left(W_{A_i}^{2n}\right)$$

$$= \prod c_i^2 \frac{(2n-1)!!}{(2n)!}\mu(A_i)^n$$

$$= \sum c_i^{2n} \frac{1}{n!} \frac{\mu(A_i)^n}{2^n}$$

$$= e^{\frac{1}{2}\sum_i c_i^2 \mu(A_i)} \to e^{\frac{1}{2}\|f\|^2}.$$

Also see Lemma 3.5. □

Example 3.11 (Brownian motion in an interval). Let $\mathscr{H} = L^2([0,1],\lambda)$, $\lambda = dx =$ the Lebesgue measure. Consider the standard Brownian motion $\{B_t\}_{t\in[0,1]}$, $B_0 = 0$, $\mathbb{E}(B_s B_t) = s \wedge t$. In this case, $W(\chi_{[0,t]}) = B_t$.

It follows that $\mathbb{E}((B_t - B_s)^2) = |t - s|$, in particular, $\mathbb{E}(B_t^2) = t$. Moreover, one has

$$\mathbb{E}(e^{B_t}) = e^{t/2}$$

$$\mathbb{E}(e^{B_t} e^{B_s}) = e^{(t+s+2t\wedge s)/2} = e^{t\wedge s} e^{(t+s)/2}$$

$$\mathbb{E}(e^{iB_t}) = e^{-t/2}$$

$$\mathbb{E}(e^{iB_t} e^{-iB_s}) = \mathbb{E}(e^{i(B_t - B_s)}) = e^{-|t-s|/2}.$$

So we must work with complex functions in $L^2(\Omega, \mathbb{P})$, which is isomorphic to the symmetric Fock space $\Gamma_{sym}(L^2[0,1])$.

Lemma 3.12 (A stochastic trick). *Set* $X_t = \exp(-t/2 + B_t)$, *where* B_t *is the usual Brownian motion,* $\mathbb{E}(B_s B_t) = s \wedge t$, *then*

$$\mathbb{E}(X_t X_s) = e^{s\wedge t}.$$

Proof. Note that

$$\mathbb{E}(e^{B_s} e^{B_t}) = \mathbb{E}(e^{B_s + B_t}) = e^{t\wedge s} e^{(t+s)/2},$$

and so

$$\mathbb{E}(\underbrace{e^{-s/2+B_s}}_{X_s} \underbrace{e^{-t/2+B_t}}_{X_t}) = e^{s\wedge t}. \tag{3.25}$$

□

More generally, if $\mathscr{H}$ is a real Hilbert space, there is a Gaussian Hilbert space $L^2(\Omega, \mathbb{P})$, containing $\{\Phi(h)\}_{h\in\mathscr{H}}$ as a dense subspace. Then

$$\varepsilon(h) := e^h = \sum_0^\infty \frac{h^{\otimes n}}{\sqrt{n!}} \xrightarrow{Z} e^{\Phi(h)-\frac{1}{2}\|h\|^2}$$

extends to an isometric isomorphism

$$Z : \Gamma_{sym}(\mathscr{H}) \to L^2(\Omega, \mathbb{P}).$$

(See Lemma 3.13 for details.)

Indeed, in this case we have

$$\mathbb{E}\Big(\underbrace{e^{\Phi(h) - \frac{1}{2}\|h\|^2}}_{Z(h)} \underbrace{e^{\Phi(k) - \frac{1}{2}\|k\|^2}}_{Z(k)}\Big) = \mathbb{E}\big(e^{\Phi(h+k)}\big) e^{-\frac{\|h\|^2 + \|k\|^2}{2}}$$

$$= e^{\langle h, k \rangle} = \langle \varepsilon(h), \varepsilon(k) \rangle_{\Gamma_{sym}}. \tag{3.26}$$

One should compare (3.26) with the simple version (3.25).

Lemma 3.13. *Let $\mathscr{H}$ be a real Hilbert space. Suppose $(\Omega, \mathscr{C}, \mathbb{P})$ is the probability space which realizes $\{\Phi(h) : h \in \mathscr{H}\}$ as a Gaussian process. Then the following sets $\{e^{\Phi(h)} : h \in \mathscr{H}\}$, and $\{e^{i\Phi(h)} : h \in \mathscr{H}\}$, are total in $L^2(\Omega, \mathbb{P})$, i.e., the closed span is all of $L^2(\Omega, \mathbb{P})$.*

Proof. We shall illustrate the second case. We now show that the family of complex exponentials of the Gaussian field is total; i.e., that the span is dense in $L^2(\mathbb{P})$. Recall, the $L^2(\mathbb{P})$-inner product is expressed as the expectation $\mathbb{E}$ of a product.

Suppose $\psi \in L^2(\Omega, \mathbb{P})$, and

$$\mathbb{E}\left(e^{-i\Phi(h)}\psi\right) = 0 \tag{3.27}$$

for all $h \in \mathscr{H}$. Fix n, $\{c_i\}_1^n$, $\{h_i\}_1^n$, $c_i \in \mathbb{R}$, $h_i \in \mathscr{H}$, set $h = \sum_{j=1}^n c_j h_j$. Let $g_n(x)$, $x \in \mathbb{R}^n$, be the Gaussian density corresponding to the Gramian $G_n = (\langle h_i, h_j \rangle)_{ij=1}^n$ as covariance matrix. (See Section 3.3.3 for more details.) Let $\mathbb{E}_n$ denote the conditional expectation onto the sub-σ-algebra $\mathscr{C}_n = \mathscr{C}(h_1, h_2, \ldots, h_n)$ determined by the cylinder sets.

From (3.27), we then get

$$\int_{\mathbb{R}^n} e^{-i\sum_{i=1}^n c_i x_i} \mathbb{E}_n(\psi)(x) g_n(x) \, dx = 0$$

for all $(c_i) \in \mathbb{R}^n$. This is a statement about vanishing Fourier transforms; and so we conclude that $\mathbb{E}_n(\psi) = 0$ in $L^2(\Omega, \mathbb{P})$. Since this holds for all such conditional expectations $\mathbb{E}_n$, then $\psi = 0$ in $L^2(\Omega, \mathbb{P})$, which is the assertion. $\square$

3.3.1 Transforms induced by isometries $V : \mathscr{H}(K) \to L^2(\mathbb{P})$ for different choices of p.d. kernels

Now, we consider the three transforms from Example 3.11 and Lemma 3.12. The corresponding p.d. kernels, and transforms, are listed in (i)–(iii) below.

For $i = 1, 2, 3$, we shall consider three transforms induced by the isometries $V_i : \mathscr{H}(K_i) \to L^2(\mathbb{P}_i)$, where $L^2(\mathbb{P}_i)$ are the corresponding Gaussian Hilbert spaces:

$$\mathscr{H}(K_i) \underset{J_i = V_i^*}{\overset{V_i}{\rightleftarrows}} L^2(\mathbb{P}_i)$$

(i) $K_1(s,t) = e^{-|s-t|/2}$, $V_1 : e^{iB_s} \longrightarrow e^{-|s-\cdot|/2} \in \mathscr{H}(K_1)$, and

$$\mathbb{E}\left(f e^{-iB_s}\right) = \langle V_1^* f, K_1(s, \cdot)\rangle_{\mathscr{H}(K_1)} = (V_1^* f)(s).$$

(ii) $K_2(s,t) = e^{s \wedge t}$, and

$$\mathbb{E}\left(e^{-s/2+B_s} f\right) = \langle V_2^* f, K_2(s, \cdot)\rangle_{\mathscr{H}(K_2)} = (V_2^* f)(s).$$

(iii) $K_3(h,k) = e^{\langle h,k\rangle}$, $h, k \in \mathscr{H}$, and

$$\mathbb{E}\left(e^{-\frac{1}{2}\|h\|^2 + \Phi(h)} f\right) = \langle V_3^* f, K_3(h, \cdot)\rangle_{\mathscr{H}(K_3)} = (V_3^* f)(s).$$

In what follows, we shall set $J_i = V_i^*$, $i = 1, 2, 3$.

3.3.2 The case $K(A, B) = \mu(A \cap B)$

Let $(M, \mathscr{B}, \mu)$ be a σ-finite measure space. Consider the p.d. kernel

$$K(A, B) := \mu(A \cap B), \tag{3.28}$$

and let $\mathscr{H}(K)$ be the corresponding RKHS.

Theorem 3.14. *$\mathscr{H}(K)$ consists of all signed measures F on $\mathscr{B}$ satisfying $F \ll \mu$ (absolutely continuous). Let $f = dF/d\mu$ be the Radon–Nikodym derivative, then*

$$\|F\|^2_{\mathscr{H}(K)} = \int |f|^2 \, d\mu. \tag{3.29}$$

Proof. With the definition of $\langle \cdot, \cdot \rangle_{\mathscr{H}(K)}$, we must show that $\langle F, \mu(A \cap \cdot)\rangle_{\mathscr{H}(K)} = F(A)$, for all $A \in \mathscr{B}$. But we have the $L^2(\mu)$-inner product and the Radon–Nikodym derivative $\chi_A\left(\cdot\right) = d\mu\left(A\cap\cdot\right)/d\mu$, so

$$\langle F, \underbrace{\mu\left(A\cap\cdot\right)}_{K(A,\cdot)} \rangle_{\mathscr{H}(K)} = \int f\chi_A d\mu = \int_A f d\mu = F\left(A\right),$$

where we used $f = dF/d\mu$ in the last step. $\qquad\square$

Corollary 3.15. $\mathscr{H}\left(K\right) \simeq L^2\left(\mu\right)$ *via* (3.29).

When passing to the Wiener process $W^{(\mu)}$ (see, e.g., Section 3.3), one gets a probability space $(\Omega, \mathscr{C}, \mathbb{P})$ where $\mathbb{P}$ depends on μ. Hence the process $\{W_A^{(\mu)}\}_{A\in\mathscr{B}_{fin}}$, $\mathscr{B}_{fin} = \{A \in \mathscr{B} : \mu\left(A\right) < \infty\}$, has its covariance kernel

$$\mathbb{E}\left(W_A^{(\mu)}W_B^{(\mu)}\right) = K(A,B) = \mu(A\cap B).$$

The mapping $K\left(A,\cdot\right) \to W_A^{(\mu)}$ extends to an isometry $V : \mathscr{H}\left(K\right) \to L^2\left(\Omega, \mathbb{P}\right)$. Moreover,

$$W\left(f\right) = \int_M f\left(x\right) dW_x^{(\mu)}, \quad f \in L^2(\mu)$$

is the Itô-integral; $f = dF/d\mu$. In particular,

$$\mathbb{E}(|W(f)|^2) = \int |f|^2 d\mu = \|F\|^2_{\mathscr{H}(K)}.$$

The analogous consideration for the p.d. kernel

$$Q\left(A,B\right) = \exp\left(\mu\left(A\cap B\right)\right), \ A, B \in \mathscr{B}_{fin}$$

is that

$$\mathbb{E}\left(e^{W_A-\frac{1}{2}\mu(A)}e^{W_B-\frac{1}{2}\mu(B)}\right) = Q(A,B),$$

$$\mathbb{E}\left(e^{W_f-\frac{1}{2}\|f\|^2}e^{W_g-\frac{1}{2}\|g\|^2}\right) = \exp(\langle f,g\rangle_{L^2}).$$

Recall that

$$\exp\left(\langle f,g\rangle_{L^2}\right) = \langle\varepsilon(f),\varepsilon(g)\rangle_{\Gamma_{sym}},$$

where $\varepsilon\left(f\right) = \sum_{n=0}^{\infty} f^{\otimes n}/\sqrt{n!}$, and Γ_{sym} denotes the symmetric Fock space over $L^2\left(\mu\right)$.

For $\psi \in L^2\left(\Omega, \mathbb{P}\right)$, we have

$$W^*\left(\psi\right)\left(f\right) = \mathbb{E}\left(\psi e^{W_f-\frac{1}{2}\|f\|^2}\right).$$

3.3.3 *Jointly Gaussian distributions*

We have stressed the existence of a centered Gaussian process, say $W(h)$ indexed by a Hilbert space $\mathscr{H}$, such that the covariance is given by the inner product from $\mathscr{H}$. But the other requirement is that the process being *jointly Gaussian*. Below we add some details about the joint distributions. For more details, see also Definition 4.103 and Figure 4.8.

Definition 3.16. A probability distribution π on $\mathbb{R}^n$ with the Borel sigma-algebra is said to be the joint distribution for the n-dimensional random vector $X = (X_1, \dots, X_n)$ if, for all Borel function $\psi : \mathbb{R}^n \to \mathbb{R}$,

$$\mathbb{E}\left(\psi\left(X\right)\right) = \int_{\mathbb{R}^n} \psi d\pi.$$

A random process (X_s) is said to be Gaussian if, for all $(s_i)_1^n$, $(X_{s_i})_1^n$ satisfies $\mathbb{E}(X_{s_i}) = 0$ and $\pi = g_n(x)\,dx$, where g_n is as in (3.30).

Specifically, we say $\{W(h) : h \in \mathscr{H}\}$ is Gaussian, if $(W(h_1), \dots, W(h_n))$ is jointly Gaussian, i.e.,

$$\mathrm{dist}\left(W(h_1), \dots, W(h_n)\right) \sim g_n(x) = \frac{1}{(2\pi)^{n/2} \det G_n} e^{-xG_n^{-1}x}, \quad x \in \mathbb{R}^n,$$

$$(3.30)$$

where $G_n = \left(\langle h_i, h_j \rangle_{\mathscr{H}}\right)_{i,j=1}^n$ is the $n \times n$ Gramian matrix, p.d., with corresponding inverse G^{-1}, and

$$xG_n^{-1}x = \sum_i \sum_j x_i \left(G_n^{-1}\right)_{ij} x_j.$$

If $A_n \in \mathscr{B}$, the Borel sigma-algebra of $\mathbb{R}^n$, then

$$\mathrm{Prob}\left(\{(W(h_1), \dots, W(h_n)) \in A\}\right) = \int_{A_n} g_n(x)\,dx,$$

and

$$\mathbb{E}\left(\psi\left(W(h_1), \dots, W(h_n)\right)\right) = \int_{\mathbb{R}^n} \psi(x)\, g_n(x)\,dx.$$

Remark 3.17. There are examples of two normal random variables X and Y, so individually Gaussian, such that the pair fails to be jointly Gaussian, and then with $X + Y$ not Gaussian. So unless the given pair of Gaussian random variables is also jointly Gaussian, the sum might fail to be Gaussian.

Example 3.18. For the standard Brownian motion $(B_t)_{t \geq 0}$, $\mathbb{E}(B_s B_t) = s \wedge t$, the joint distribution of $(B_{t_1}, \ldots, B_{t_n})$, with $t_1 < \cdots < t_n$, is determined by the Gramian matrix

$$G_n = \begin{bmatrix} t_1 & t_1 & t_1 & t_1 & \cdots & t_1 \\ t_1 & t_2 & t_2 & t_2 & \cdots & t_2 \\ t_1 & t_2 & t_3 & t_3 & \cdots & t_3 \\ \vdots & \vdots & \vdots & \ddots & \ddots & \vdots \\ t_1 & t_2 & t_3 & \vdots & t_{n-1} & t_{n-1} \\ t_1 & t_2 & t_3 & \cdots & t_{n-1} & t_n \end{bmatrix}. \tag{3.31}$$

The matrix G_n admits a factorization

$$G_n = A_n A_n^*, \tag{3.32}$$

where A_n is lower-triangular,

$$A_n = \begin{bmatrix} \sqrt{t_1} & 0 & 0 & \cdots & 0 \\ \sqrt{t_1} & \sqrt{t_2 - t_1} & 0 & \cdots & \vdots \\ \sqrt{t_1} & \sqrt{t_2 - t_1} & \sqrt{t_3 - t_2} & \ddots & \vdots \\ \vdots & \vdots & \vdots & \ddots & 0 \\ \sqrt{t_1} & \sqrt{t_2 - t_1} & \sqrt{t_3 - t_2} & \cdots & \sqrt{t_n - t_{n-1}} \end{bmatrix}. \tag{3.33}$$

In particular,

$$\det(G_n) = t_1 (t_2 - t_1) \cdots (t_N - t_{n-1}). \tag{3.34}$$

3.3.4 *CND kernels*

While conditionally negative definite (CND) functions play an important role in harmonic analysis. For those applications the context may be more general, e.g., complex valued.

Here we shall restrict the setting to that motivated by work of Schoenberg and von Neumann (see especially [Sch37].) This in turn is what is involved in our present applications to Gaussian processes.

Definition 3.19. Let S be a set. A function $N : S \times S \to \mathbb{R}$ is *conditionally negative definite*, if for all finite systems $(c_i)_{i=1}^n$, $(s_i)_{i=1}^n$, $c_i \in \mathbb{R}$, $\sum c_i = 0$, $s_i \in S$, we have

$$\sum_i \sum_j c_i c_j N(s_i, s_j) \leq 0.$$

Theorem 3.20 ([Sch37]). *Let S be a set, $N : S \times S \to \mathbb{R}$ CND. Then there exists a Hilbert space $\mathscr{H}$, and a mapping $S \ni s \to \lambda(s) \in \mathscr{H}$, such that*

$$N(s,t) = \|\lambda(s) - \lambda(t)\|_{\mathscr{H}}^2.$$

We shall return to further analysis of conditionally negative definite functions (CND), with applications, in Chapter 4; see especially the material following Theorem 4.22.

Corollary 3.21. *If N is a CND function, then $e^{-\frac{1}{2}N(s,t)}$ is positive definite.*

Proof. Note that

$$e^{-\frac{1}{2}N(s,t)} = e^{-\frac{1}{2}\|\lambda(s)-\lambda(t)\|_{\mathscr{H}}^2}$$
$$= e^{-\frac{1}{2}\|\lambda(s)\|_{\mathscr{H}}^2} e^{\langle \lambda(s),\lambda(t)\rangle_{\mathscr{H}}} e^{-\frac{1}{2}\|\lambda(t)\|_{\mathscr{H}}^2}.$$

The exponential in the middle of this product is clearly p.d., see also Lemma 3.4 and details below. The product of the remaining two factors (the first and the last) constitutes a rank-one p.d. function. And so the conclusion in the corollary follows since the product of any pair of p.d. functions is again p.d.

Recall that

$$e^{\langle \lambda(s),\lambda(t)\rangle_{\mathscr{H}}} = \left\langle \sum \frac{\lambda(s)^{\otimes n}}{\sqrt{n!}}, \sum \frac{\lambda(t)^{\otimes n}}{\sqrt{n!}} \right\rangle_{\Gamma_{sym}(\mathscr{H})}$$

by the computation with a symmetric Fock space (Section 3.2), where

$$\Gamma_{sym}(\mathscr{H}) = \mathbb{C} \oplus \mathscr{H} \oplus \mathscr{H}^{\otimes 2} \oplus \cdots. \qquad \square$$

Remark 3.22. Comparison of kernels K which are positive definite (p.d.) vs the case of conditionally negative definite (CND), designated N.

Below we outline that both classes of kernels lead to Hilbert completions, $\mathscr{H}(K)$, respectively $\mathscr{H}(N)$. The first one is the familiar RKHS, while the second is different. In summary; the key property for the case of $\mathscr{H}(K)$ is that the values $f(s)$ for f in $\mathscr{H}(K)$ are reproduced by $K(s,\cdot)$. By contrast, for $\mathscr{H}(N)$ and f in $\mathscr{H}(N)$, it is the set of differences $f(s)-f(t)$ which are reproduced. The property is called "relative reproducing."

(1) p.d. kernels K

$$V_K := \left\{ \sum_i c_i K(s_i,\cdot) \right\} \xrightarrow[\text{completion}]{\text{Hilbert}} \mathscr{H}(K)$$

$$\left\| \sum_i c_i K\left(s_i, \cdot\right) \right\|^2_{\mathscr{H}(K)} = \sum_i \sum_j c_i c_j K\left(s_i, s_j\right)$$

$$\langle f, K\left(s, \cdot\right) \rangle_{\mathscr{H}(K)} = f(s), \quad \forall f \in \mathscr{H}\left(K\right)$$

In particular,

$$\| K\left(s, \cdot\right) \|^2_{\mathscr{H}(K)} = K(s, s).$$

(2) CND kernels N

$$V_N := \left\{ \sum_i c_i N\left(s_i, \cdot\right) : \sum_i c_i = 0 \right\} \xrightarrow[\text{completion}]{\text{Hilbert}} \mathscr{H}\left(N\right)$$

$$\left\| \sum_i c_i N\left(s_i, \cdot\right) \right\|^2_{\mathscr{H}(N)} = -\frac{1}{2} \sum_i \sum_j c_i c_j N\left(s_i, s_j\right)$$

$$\langle f, N\left(s, \cdot\right) - N\left(t, \cdot\right) \rangle_{\mathscr{H}(N)} = f(s) - f(t), \quad \forall f \in \mathscr{H}(N).$$

When taking the Hilbert completion, one divides out functions

$$g\left(\cdot\right) = \sum_i c_i N(s_i, \cdot), \quad \sum_i c_i = 0.$$

In particular,

$$\| N\left(s, \cdot\right) - N\left(t, \cdot\right) \|^2_{\mathscr{H}(N)} = -\frac{1}{2} \big[\underbrace{N\left(s, s\right)}_{=0} + \underbrace{N\left(t, t\right)}_{=0} - 2N\left(s, t\right) \big]$$

$$= N(s, t).$$

Theorem 3.23. *Let S be a general set. Let $\lambda : S \to \mathscr{H}\left(\lambda\right) \subset l^2\left(\mathbb{N}\right)$ be 1–1, where $\mathscr{H}\left(\lambda\right)$ is a closed subspace in $l^2\left(\mathbb{N}\right)$. Set*

$$K\left(s, t\right) = \langle \lambda(s), \lambda(t) \rangle_{l^2}, \quad s, t \in S;$$

then

$$d\left(s, t\right)^2 = \| \lambda\left(s\right) - \lambda\left(t\right) \|^2_{l^2} =: N\left(s, t\right)$$

determines the K-metric.

We may also write

$$K\left(s, t\right) = \sum_n e_n\left(s\right) e_n\left(t\right), \tag{3.35}$$

where $\lambda\left(s\right) = \left(e_n\left(s\right)\right)_{n \in \mathbb{N}}$. If $\{b_n\}$ is an ONB for $\mathscr{H}(\lambda)$, then

$$e_n\left(s\right) = \langle \lambda(s), b_n \rangle_{l^2}. \tag{3.36}$$

Lemma 3.24. *The representation* (3.35) *is independent of choices of ONBs, see* (3.36).

Proof. If we pick two ONBs (b_n) and (b'_n) for $\mathscr{H}(\lambda)$, then there exists a unitary operator U in $\mathscr{H}(\lambda)$ such that $Ub_n = b'_n$, and so we get

$$
\begin{aligned}
\sum_n \langle \lambda(s), b'_n \rangle_{l^2} \langle \lambda(t), b'_n \rangle_{l^2} &= \sum_n \langle \lambda(s), Ub_n \rangle_{l^2} \langle \lambda(t), Ub_n \rangle_{l^2} \\
&= \sum_n \langle U^* \lambda(s), b_n \rangle_{l^2} \langle U^* \lambda(t), b_n \rangle_{l^2} \\
&= \langle U^* \lambda(s), U^* \lambda(t) \rangle_{l^2} \\
&= \langle \lambda(s), \lambda(t) \rangle_{l^2},
\end{aligned}
$$

which proves the assertion. $\qquad\square$

Corollary 3.25. *Let S be a set, and let d be a metric on S. Then the metric d has the form $d = d_K$ for a p.d. function K on $S \times S$ if and only if*

$$
(s, t) \longmapsto d(s, t)^2
$$

is conditionally negative definite (CND).

Proof. This follows from the discussion above, and from Schoenberg's theorem (CND): $\forall n$, $\forall (c_i)$, $\forall (s_i)$, $c_i \in \mathbb{R}$, $s_i \in S$ such that $\sum_i c_i = 0$, we have $\sum_i \sum_j c_i c_j d(s_i, s_j)^2 \le 0$. $\qquad\square$

3.4 Guide to the literature

While we have cited some paper from the Bibliography in our discussion inside the chapter, readers not familiar with the ideas involved may wish to consult the following sources: [BR81, GJ87, JT17, JP91, LG14, PS72a, PS11, vN31, Jan97, BCR84].

Chapter 4

Gaussian Stochastic Processes: Gaussian Fields and Their Realizations

"Statistics is the grammar of science." Karl Pearson.

"While writing my book [Stochastic Processes] I had an argument with Feller. He asserted that everyone said 'random variable' and I asserted that everyone said 'chance variable.' We obviously had to use the same name in our books, so we decided the issue by a stochastic procedure. That is, we tossed for it and he won." Doob, J. Quoted in Statistical Science.

While the subject Gaussian processes, stochastic calculus, and probabilistic structures is vast, with many and varied subdisciplines, and with many and diverse applications, for our present purposes we have made choices as follows, each motivated by the aim of the book: The present selection of topics is primarily motivated by its use in our presentation of Malliavin calculus. Useful supplement treatments in the literature: [BCR84, Hid03, Hid07, Itô04, Itô06, Mal78].

In the previous chapter we discussed representations of the CCRs (among other things). The non-commutativity from this setting forces a study of commutators. We shall show below that it also forces an important infinite dimensional calculus. The important terms from this are Gaussian processes, Gaussian fields, It calculus, Malliavin derivatives, stochastic differential equations, and stochastic PDEs.

By a Gaussian process (also called Gaussian fields), we mean a system of random variables, such that all joint distributions are Gaussian. We usually normalize, so reducing to the case of mean zero. We shall study a fruitful interplay between the following three closely related settings: (i) Gaussian processes indexed by a set S; (ii) positive definite (p.d.) kernels K defined on $S \times S$; and (iii) a certain class of representations. The connection between (i) and (ii) can be summarized as follows: Every p.d kernel on $S \times S$ is the covariance kernel for a Gaussian process indexed by S; see Section 4.1.1 below for details.

4.1 Analysis on reproducing Kernel Hilbert space (RKHS)

The theory of *reproducing kernel Hilbert space* (RKHS) is usually credited to N. Aronszajn [Aro50, Aro61], see especially his 1950 paper in TAMS; but many others contributed to a variety of aspects [AAH69, AS57, PR16, Nel58, JT19b, JT19a, JT18b, BSS90, BSS89]. Also see [AESW51, Sch42, vNS41, Sch38, Sch37].

Definition 4.1. A RKHS is a Hilbert space $\mathscr{H}$ of functions on some set S such that for $\forall s \in S$, $\mathscr{H} \ni f \to f(s)$ is continuous in the norm $\|\cdot\|$ of $\mathscr{H}$.

Lemma 4.2. *Given the data, use Riesz (Lemma 1.18), one has* $f(s) = \langle f, K_s(\cdot) \rangle$, $f \in \mathscr{H}$, $s \in S$, $K_s \in \mathscr{H}$. *Let* $K(s,t) = \langle K_s(\cdot), K_t(\cdot) \rangle_{\mathscr{H}}$, *then* K *is p.d.*

Corollary 4.3. *If* $K : S \times S \to \mathbb{R}$ *is specified positive definite (p.d.), then there exists a unique* $\mathscr{H} = \mathscr{H}(K)$ *which is a RKHS with* K *as its p.d. kernel.*

Proof. The standard literature; see, e.g., [Aro50, Jan97] and Section 1.1.1.
$\square$

With the setting as above, and starting with a fixed p.d. kernel K on $S \times S$, we then arrive at the corresponding RKHS $\mathscr{H}(K)$. Naturally, this is a space of functions on S, but we will need a *precise condition* which allows us to decide the question of whether a given function F, defined on S, is in $\mathscr{H}(K)$, or not. Off hand, deciding the question is not immediate since $\mathscr{H}(K)$, by its construction, is only specified *as an abstract Hilbert completion*. Now the same question also arises, in different contexts, later in the present chapter. But for our immediate purpose, Lemma 4.4 here will suffice. Below, we therefore outline the main idea, and we shall then return and offer more context, and complete proof details later in Lemmas 4.52, and 4.64.

Lemma 4.4. *Let* $K : S \times S \to \mathbb{R}$ *be a p.d. kernel, and* $\mathscr{H}(K)$ *the corresponding RKHS. Let* F *be a function on* S. *Then the following are equivalent:*

(i) $F \in \mathscr{H}(K)$;

(ii) *For all* $n \in \mathbb{N}$, *for all* $(c_i)_{i=1}^n$ *and* $(s_i)_{i=1}^n$ *with* $c_i \in \mathbb{R}$, $s_i \in S$, *there exists a constant* C_F *such that*

$$\left| \sum c_i F(s_i) \right|^2 \leq C_F \sum_i \sum_j c_i c_j K(s_i, s_j). \tag{4.1}$$

If $F \in \mathscr{H}(K)$, *then* $\|F\|_{\mathscr{H}(K)}^2$ *is equal to the least constant* C_F *for which* (4.1) *holds.*

Proof. Given (i), i.e., $F \in \mathscr{H}(K)$, then by Cauchy–Schwarz, we get that

$$\left| \sum c_i F(s_i) \right|^2 = \left| \left\langle F, \sum c_i K(\cdot, s_i) \right\rangle_{\mathscr{H}(K)} \right|^2$$

$$\leq \|F\|_{\mathscr{H}(K)}^2 \left\| \sum c_i K(\cdot, s_i) \right\|_{\mathscr{H}(K)}^2$$

$$= C_F \sum_i \sum_j c_i c_j K(s_i, s_j),$$

and so $C_F = \|F\|_{\mathscr{H}(K)}^2$.

Conversely, suppose (4.1) holds. Since the set $\{\sum c_i K(s_i, \cdot) :$ finite sums$\}$ is dense in $\mathscr{H}(K)$, then the mapping

$$l : \sum c_i K(s_i, \cdot) \longmapsto \sum c_k F(s_i)$$

extends to a bounded linear functional on $\mathscr{H}(K)$, with norm $\|l\|$ bounded by $C_F^{1/2}$. It follows from Lemma 1.18, that there exists a unique $G_l \in \mathscr{H}(K)$ such that

$$\sum c_k F(x_i) = \left\langle G_l, \sum c_i K(x_i, \cdot) \right\rangle_{\mathscr{H}(K)}.$$

In particular,

$$F(x) = \langle G_l, K(x) \rangle_{\mathscr{H}(K)} = G(x),$$

for all $x \in S$. The last step in the above calculation follows from the reproducing property. $\qquad\square$

Consider a p.d. kernel $K : S \times S \to \mathbb{R}$. Then define a realization $\rho : S \to \mathcal{H}(K)$ specified by $\rho(s) = K(s, \cdot)$. Note that then

$$\langle \rho(s), \rho(t) \rangle_{\mathcal{H}(K)} = K(s, t). \tag{4.2}$$

Lemma 4.5. *For any pair $(\psi, \mathcal{L})$ where $\mathcal{L}$ is a Hilbert space, and $\psi : S \to \mathcal{L}$ is a function with $K(s, t) = \langle \psi(s), \psi(t) \rangle_{\mathcal{L}}$, there is a unique isometry $W : \mathcal{H}(K) \to \mathcal{L}$ such that $W \circ \rho = \psi$; see the diagram below.*

$$\tag{4.3}$$

Proof. Let $(\psi, \mathcal{L})$ be as specified subject to condition (4.2). For $s \in S$, set $W(K_s) := \psi(s)$, and extend W by linearity, i.e.,

$$W\left(\sum_i c_i K_{s_i}\right) := \sum_i c_i \psi(s_i) \tag{4.4}$$

with finite sums. Then W has the desired properties. To see that, it is enough to verify the *isometry* property on the dense space $\{\sum_i c_i K_{s_i}\}$:

$$\left\| W\left(\sum_i c_i K_{s_i}\right) \right\|_{\mathcal{L}}^2 = \left\| \sum_i c_i \psi(s_i) \right\|_{\mathcal{L}}^2$$

$$= \sum_i \sum_j c_i c_j \langle \psi(s_i), \psi(s_j) \rangle_{\mathcal{L}}$$

$$= \sum_i \sum_j c_i c_j K(s_i, s_j) \quad \text{by (4.2)}$$

$$= \left\| \sum c_i K_{s_i} \right\|_{\mathcal{H}(K)}^2 \quad \text{by the RKHS axiom,}$$

which is the desired conclusion. $\qquad\square$

Corollary 4.6. *Let K and L be two positive definite functions defined on $S \times S$, and set $M(s, t) = K(s, t)L(s, t)$, for all $s, t \in S$; then M is also p.d., and for their respective RKHSs we have:*

$$\mathcal{H}(M) \simeq \mathcal{H}(K) \otimes \mathcal{H}(L) \tag{4.5}$$

where the RHS in (4.5) denotes the tensor product of Hilbert spaces.

Remark 4.7. In (4.5), we state the conclusion up to isomorphism of Hilbert spaces. For example, if $K(z, w) = \frac{1}{1-z\overline{w}}$, $z, w \in \mathbb{D} = \{z \in \mathbb{C} : |z| < 1\}$, then the RKHS for $K^2(z, w) = \frac{1}{(1-z\overline{w})^2}$ is the Bergman Hilbert space of (analytic functions on $\mathbb{D}$) $\cap\ L^2(\mathbb{D}, A)$, where A is the plane measure on $\mathbb{D}$.

Let $\mathscr{L}$ be a real (or complex) Hilbert space. We showed that $\mathscr{L}$ always has an orthogonal basis (ONB) $(e_\alpha)_{\alpha \in I}$, indexed by some set I. If $(e_\alpha)_{\alpha \in I}$ and $(f_\beta)_{\beta \in J}$ are two ONBs, then the sets I and J have the same cardinality. If $card(I) = \aleph_0 = card(\mathbb{N})$, we say that the Hilbert space L is separable; and we shall restrict attention to the separate case.

Corollary 4.8. *Let $K : S \times S \to \mathbb{R}$ be a p.d. kernel, and let $(e_n)_{n \in \mathbb{N}}$ be a system of functions on S such that*

$$K(s, t) = \sum_{n \in \mathbb{N}} e_n(s)\, e_n(t) \tag{4.6}$$

holds. Then:

(i) $e_n(\cdot) \in \mathscr{H}(K)$, *and* $\|e_n\|_{\mathscr{H}} \leq 1$, *for all* $n \in \mathbb{N}$.

(ii) $\{e_n\}$ *is a Parseval frame (see Definition 1.16) in $\mathscr{H}(K)$. Specifically, for all $F \in \mathscr{H}(K)$, we have:*

$$F(t) = \sum_{n \in \mathbb{N}} \langle F, e_n \rangle_{\mathscr{H}(K)}\, e_n(t), \tag{4.7}$$

with

$$\|F\|^2_{\mathscr{H}(K)} = \sum_{n=1}^{\infty} \left| \langle F, e_n \rangle_{\mathscr{H}(K)} \right|^2. \tag{4.8}$$

(iii) $\mathscr{H}(K)$ *consists of functions F on S of the form*

$$F(s) = \sum_{n \in \mathbb{N}} \alpha_n e_n(s) \tag{4.9}$$

where $(\alpha_n) \in l^2(\mathbb{N})$. However, the coefficients in (4.9) are not unique unless (e_n) is an ONB.

(iv) $\{e_n\}$ *is an ONB if and only if $\|e_n\|_{\mathscr{H}(K)} = 1$ for all n. Then the mapping $V : \mathscr{H}(K) \to l^2$ defined as*

$$F \longmapsto VF = \left(\langle F, e_n \rangle_{\mathscr{H}(K)} \right) \tag{4.10}$$

is an isometric isomorphism. (In particular, V is onto.)

Proof. (i) Fix n, for example $n = 1$, then for all finite systems (c_i), (s_i), $c_i \in \mathbb{R}$, $s_i \in S$, we have

$$\left| \sum c_i e_1 (s_i) \right|^2 = \sum_i \sum_j c_i c_j e_1 (s_i) e_1 (s_j)$$

$$\leq \sum_{n \in \mathbb{N}} \sum_i \sum_j c_i c_j e_n (s_i) e_n (s_j)$$

$$= \sum_i \sum_j c_i c_j K (s_i, s_j).$$

Therefore, by Lemma 4.4, we conclude that $e_1 \in \mathscr{H}(K)$ and $\|e_1\|_{\mathscr{H}(K)} \leq 1$.
(ii) By the reproducing property, we have

$$F(t) = \langle F, K(\cdot, t) \rangle_{\mathscr{H}(K)}$$

$$= \left\langle F, \sum_{n \in \mathbb{N}} e_n(\cdot) e_n(t) \right\rangle_{\mathscr{H}(K)}$$

$$= \sum_{n \in \mathbb{N}} \langle F, e_n \rangle_{\mathscr{H}(K)} e_n(t)$$

which is (4.7). Now,

$$\|F\|^2_{\mathscr{H}(K)} = \langle F, F \rangle_{\mathscr{H}(K)}$$

$$= \sum_{n \in \mathbb{N}} \langle F, e_n \rangle_{\mathscr{H}(K)} \langle e_n(\cdot), F \rangle_{\mathscr{H}(K)}$$

$$= \sum_{n \in \mathbb{N}} \left| \langle F, e_n \rangle_{\mathscr{H}(K)} \right|^2,$$

which is the desired conclusion (4.8).
(iii) Note that

$$K(s, s) = \sum_n e_n(s)^2 < \infty,$$

and so $(e_n(s)) \in l^2$, for all $s \in S$. By (4.7) & (4.8), we see that every $F \in \mathscr{H}(K)$ has an expansion as in (4.9) with coefficients in l^2.
(iv) For a fixed n, we have

$$\|e_n\|^2_{\mathscr{H}(K)} = \sum_{m \in \mathbb{N}} \left| \langle e_n, e_m \rangle_{\mathscr{H}(K)} \right|^2$$

$$= \|e_n\|^2_{\mathscr{H}(\mathscr{K})} + \sum_{m \neq n} \left| \langle e_n, e_m \rangle_{\mathscr{H}(K)} \right|^2. \tag{4.11}$$

Substitute $\|e_n\|_{\mathscr{H}(k)} = 1$ into (4.11), and we conclude that $\langle e_n, e_m \rangle_{\mathscr{H}(K)} = 0$ if $m \neq n$, which is the orthogonality. $\square$

Remark 4.9.

(i) Assume (4.6). We get from conclusion (i) that the functions $e_n(\cdot)$ in the system will automatically be Lipschitz continuous in a natural metric on S.

(ii) The representation (4.6) is a special case of measure factorization

$$K(s,t) = \int_X \psi(x,s)\, \psi(x,t)\, \mu(dx)$$

over a measure space $(X, \mathscr{B}_X, \mu)$. Equation (4.6) corresponds to $X = \mathbb{N}$, and $\mu = $ the counting measure. We then conclude that $\psi(\cdot, s) \in L^2(\mu)$ for all $s \in S$, and for all $A \in \mathscr{B}_X$ with finite measure, we have

$$s \longmapsto \int_A \psi(x,s)\, \mu(ds) \in \mathscr{H}(K).$$

To see this, note the argument is similar to that in the proof of part (i), but now

$$\left| \sum_i c_i \int_A \psi(x, s_i)\, \mu(dx) \right|^2 \leq \mu(A) \sum_i \sum_j c_i c_j K(s_i, s_j).$$

The assertion again follows from Lemma 4.4.

(iii) In part (iv), in order to deduce that $\{e_n\}$ is an ONB, we must assume $\|e_n\|_{\mathscr{H}(K)} = 1$ for *every* n. For example, if (e_n) denotes the standard ONB in $l^2(\mathbb{N})$, then

$$e_1, e_2/\sqrt{2}, e_2/\sqrt{2}, e_3, e_4, e_5, \ldots$$

is a Parseval frame in $l^2(\mathbb{N})$. Note that $\|e_1\|_{l^2} = 1$, but the other vectors, aside from the first one, are not mutually orthogonal.

Corollary 4.10. *Assume (4.6), i.e.,*

$$K(s,t) = \sum e_n(s)\, e_n(t).$$

Define $V : \mathscr{H}(K) \to l^2(\mathbb{N})$ by

$$VF = \left(\langle F, e_n \rangle_{\mathscr{H}(K)} \right)_n. \tag{4.12}$$

Then V is isometric, and $V^* : l^2(\mathbb{N}) \to \mathscr{H}(K)$ is given by

$$V^*((\alpha_n)) = \sum \alpha_n e_n(\cdot). \tag{4.13}$$

In particular

$$V^*V = I_{\mathscr{H}(K)}. \qquad (4.14)$$

Proof. We need only verify that the adjoint operator V^* is given by (4.13). Fix $F \in \mathscr{H}(K)$ and $(\alpha_i) \in l^2(\mathbb{N})$, then

$$\langle VF, (\alpha_i) \rangle_{l^2} = \sum \langle F, e_n \rangle_{\mathscr{H}(K)} \alpha_n$$

$$= \left\langle F, \sum \alpha_n e_n \right\rangle_{\mathscr{H}(K)}$$

$$= \langle F, V^*((\alpha_i)) \rangle_{\mathscr{H}(K)},$$

and so the desired conclusion (4.13) holds. $\qquad\square$

Remark 4.11 (uniqueness of representations). Given (4.6), we get $(e_n(s))_n \in l^2(\mathbb{N})$ for all $s \in S$, and so

$$l^2(K) := \overline{span}\{(e_n(s))_n : s \in S\} = V(\mathscr{H}(K))$$

is a subspace in $l^2(\mathbb{N})$, and

$$l^2(\mathbb{N}) \ominus l^2(K) = \left\{ (\alpha_n) \in l^2(\mathbb{N}) : \sum \alpha_n e_n(s) = 0, \forall s \in S \right\} = ker(V^*).$$

Therefore,

$$\mathscr{H}(K) \simeq l^2(\mathbb{N}) / ker(V^*).$$

Assume $\{e_n\}$ is a Parseval frame. If $ker(V^*) \neq \emptyset$, then every $F \in \mathscr{H}(K)$ has a representation as in (4.7), but it may have other representations $F(s) = \sum \alpha_n e_n(s)$ with $\alpha_n \neq \langle F, e_n \rangle_{\mathscr{H}(K)}$. In both cases, however, the following identity holds:

$$\|F\|^2_{\mathscr{H}(K)} = \sum \left| \langle F, e_n \rangle_{\mathscr{H}(K)} \right|^2 = \sum |\alpha_n|^2.$$

Remark 4.12. The two operators V and V^* in Corollary 4.10 are well defined. In general, assume (4.6), i.e., $K(s,t) = \sum e_n(s) e_n(t)$, we get $\{e_n\}$ a Parseval frame in $\mathscr{H}(K)$, but V^* may not be isometric, i.e., $VV^* \neq I_{l^2}$.

In fact, if (δ_n) denotes the standard ONB in $l^2(\mathbb{N})$, then $V^*(\delta_n) = e_n$. If $VV^* = I_{l^2}$ holds, then

$$\langle V^*\delta_n, V^*\delta_m \rangle_{\mathscr{H}(K)} = \langle \delta_n, \delta_m \rangle_{l^2} = \delta_{n,m},$$

it follows that then $\{e_n\}$ must be an ONB.

Summary. Given (4.6), let V and V^* be as in (4.12) and (4.13), respectively; then $V^*V = I_{\mathscr{H}(K)}$, but $Q = VV^*$ is a non-trivial projection on $l^2(\mathbb{N})$, i.e.,

Q = the projection onto $V(\mathscr{H}(K))$, as a closed subspace in $l^2(\mathbb{N})$. See the diagram below.

$$\mathscr{H}(K) \overset{V}{\underset{V^*}{\rightleftarrows}} Q(l^2(\mathbb{N})) \subset l^2(\mathbb{N})$$

Example 4.13.

(i) $K(s,t) = \frac{1}{1-st}$, $s,t \in (-1,1)$; take $e_n(s) = s^n$.

$$\mathscr{H}(K) = \left\{ F(s) = \sum_{n=0}^{\infty} \alpha_n s^n : (\alpha_n) \in l^2(\mathbb{N}_0) \right\}.$$

(ii) $K(z,w) = \frac{1}{1-z\bar{w}}$, $z,w \in \mathbb{D} = \{z : |z| < 1\}$; take $e_n(z) = z^n$.

$$\mathscr{H}(K) = \left\{ F(z) = \sum_{n=0}^{\infty} \alpha_n z^n : (\alpha_n) \in l^2(\mathbb{N}_0) \right\}.$$

(iii) $K(s,t) = e^{st}$, $s,t \in \mathbb{R}$; take $e_n(s) = s^n / \sqrt{n!}$.

$$\mathscr{H}(K) = \left\{ F(s) = \sum_{n=0}^{\infty} \alpha_n s^n / \sqrt{n!} : (\alpha_n) \in l^2(\mathbb{N}_0) \right\}.$$

In all three cases, we have $\|F\|_{\mathscr{H}(K)}^2 = \sum_{\alpha \in \mathbb{N}_0} |\alpha_n|^2$.

The following result shows that, every p.d. kernel $K : S \times S \to \mathbb{R}$ is the covariance kernel of a Gaussian process. The complex valued case follows by an easy modification. See also Sections 4.3.1 and 4.4.

Theorem 4.14. *Given a p.d. kernel* $K : S \times S \to \mathbb{R}$, *there is a Gaussian process* $\{X_s\}_{s \in X}$ *with* $K(\cdot, \cdot)$ *as its covariance kernel. Specifically, the joint distribution of* $\{X_s\}_{s \in S}$ *is Gaussian,* $\mathbb{E}(X_s) = 0$, *for all* $s \in S$, *where* $\mathbb{E}$ *refers to the expectation defined for a probability space* $(\Omega, \mathscr{C}, \mathbb{P})$; *so* $\mathbb{E}(\cdot) = \int_{\Omega} (\cdot) \, d\mathbb{P}$. *The covariance condition then reads as follows:*

$$\mathbb{E}(X_s X_t) = K(s,t), \quad \forall s,t \in S;$$

in particular,

$$X_s \sim N(0, K(s,s)).$$

That is, the Gaussian random variable X_s *has mean zero and variance equal to* $K(s,s)$.

Proof. Pick an orthonormal basis (ONB) $\{e_n\}_{n\in\mathbb{N}}$ in $\mathscr{H}(K)$; and pick a system of i.i.d. (independent identically distributed) standard Gaussians $Z_n \sim N(0,1)$, $n \in \mathbb{N}$. Then set

$$X_s(\cdot) = \sum_{n\in\mathbb{N}} e_n(s) Z_n(\cdot), \quad s \in S. \tag{4.15}$$

By general probability theory, we have that each random variable X_s in (4.15) is Gaussian; and so we compute the covariance of the combined process $\{X_s\}_{s\in S}$:

$$\mathbb{E}(X_s X_t) = \sum_n \sum_m e_n(s) e_m(t) \mathbb{E}(Z_n Z_m)$$

$$= \sum_n \sum_m e_n(s) e_m(t) \delta_{n,m}$$

$$= \sum_n e_n(s) e_n(t) = K(s,t);$$

where we used (in the last step) that $\{e_n\}$ was chosen to be an ONB in $\mathscr{H}(K)$, so that

$$K(s,t) = \langle K_s, K_t \rangle_{\mathscr{H}(K)}$$

$$= \sum_n \langle K_s, e_n \rangle_{\mathscr{H}(K)} \langle e_n, K_t \rangle_{\mathscr{H}(K)}$$

$$= \sum_n e_n(s) e_n(t). \qquad \square$$

4.1.1 *Stochastic analysis and positive definite kernels*

Stochastic analysis is one among a diverse list of applications of positive definite kernels and the corresponding RKHSs. It turns out that the link to stochastic analysis passes through a closer study of classes of signed measures associated to the fundamental setting of positive definite kernels. Below we turn to a systematic study of these spaces of measures.

Consider a suitable transform for the more general class of positive definite function $K : S \times S \to \mathbb{R}$ (or $\mathbb{C}$), i.e., all finite sums satisfy

$$\sum_i \sum_j c_i c_j K(s_i, s_j) \geq 0,$$

where $\{c_i\}_{i=1}^n$, $\{s_i\}_{i=1}^n$, $s_i \in S$, $c_i \in \mathbb{R}$. Let $\mathscr{H}(K)$ be the corresponding RKHS.

We can introduce a σ-algebra on S generated by $\{K(\cdot, t)\}_{t \in S}$ such that $\mu \mapsto \int \mu(ds) K(s, \cdot)$ makes sense, where μ is a suitable signed measure; but it is more useful to consider linear functionals l on $\mathcal{H}(K)$.

Definition 4.15. We say that $l \in \mathcal{L}_2(K)$ if, there exists $C = C_l < \infty$ such that $|l(G)| \leq C_l \|G\|_{\mathcal{H}(K)}$, for all $G \in \mathcal{H}(K)$. On $\mathcal{L}_2(K)$, introduce the Hilbert inner product

$$lKl := \langle l, l \rangle_{\mathcal{L}_2} = l \,(\text{acting in } s) \, K(s, t) \, l \,(\text{acting in } t).$$

Definition 4.16. Given $K : S \times S \to \mathbb{R}$ p.d., let $\mathcal{B}_{(S,K)}$ be the cylinder σ-algebra on S and consider signed measures μ on $\mathcal{B}_{(S,K)}$. Let

$$\mathfrak{M}_2(K) := \left\{ \mu \mid \mu K \mu := \langle \mu, \mu \rangle_{\mathfrak{M}_2} = \iint \mu(ds) K(s, t) \mu(dt) < \infty \right\}$$

where $\langle \mu, \mu \rangle_{\mathfrak{M}_2}$ is a Hilbert pre-inner.

The basic idea with the approach using linear functionals, $l : \mathcal{H}(K) \to \mathbb{R}$, is to extend from the case of finite atomic measures to a precise completion.

Example 4.17. For $\{c_i\}_1^n$, $\{s_i\}_1^n$ $c_i \in \mathbb{R}$, let

$$l(G) := \sum_i c_i G(s_i), \quad \forall G \in \mathcal{H}(K); \tag{4.16}$$

which is the case when $\mu = \sum_i c_i \delta_{s_i}$, with δ_{s_i} denoting the Dirac measure. Setting

$$T_K(\mu) = \sum_i c_i K(s_i, \cdot) \in \mathcal{H}(K),$$

we then get

$$l(G) = \langle T_K(\mu), G \rangle_{\mathcal{H}(K)}, \quad \forall G \in \mathcal{H}(K). \tag{4.17}$$

Moreover, the following isometric property holds for all finite atomic measures μ:

$$\|T_K(\mu)\|_{\mathcal{H}(K)} = \|\mu\|_{\mathfrak{M}_2(K)}.$$

We include some basic facts about $\mathfrak{M}_2(K)$ and $\mathcal{L}_2(K)$.

Lemma 4.18.

(i) *Suppose*

$$K(s,t) = \sum_{n \in \mathbb{N}} e_n(s) e_n(t) \qquad (4.18)$$

for some functions $(e_n)_{n \in \mathbb{N}}$, *then* $\{e_n\} \subset \mathscr{H}(K)$, *and for all* $G \in \mathscr{H}(K)$,

$$\|G\|^2_{\mathscr{H}(K)} = \sum \left| \langle G, e_n \rangle_{\mathscr{H}(K)} \right|^2.$$

That is, (e_n) *is a Parseval frame in* $\mathscr{H}(K)$.

(ii) *We have*

$$l \in \mathscr{L}_2(K) \iff \sum_{n \in \mathbb{N}} |l(e_n)|^2 < \infty.$$

(iii) *Suppose* K *admits a representation (4.18). Let* l *be a linear functional on* $\mathscr{H}(K)$, *then*

$$l \in \mathscr{L}_2(K) \iff (l(e_n))_{n \in \mathbb{N}} \in l^2(\mathbb{N}).$$

Proof. Given (4.18), if $G \in \mathscr{H}(K)$, then

$$G(t) = \sum \langle G, e_n \rangle_{\mathscr{H}(K)} e_n(t), \qquad (4.19)$$

and

$$
\begin{aligned}
|l(G)|^2 &= \left| \sum \langle G, e_n \rangle_{\mathscr{H}(K)} l(e_n) \right|^2 \\
&\leq \sum \left| \langle G, e_n \rangle_{\mathscr{H}(K)} \right|^2 \sum |l(e_n)|^2 \\
&\leq C_l \|G\|^2_{\mathscr{H}(K)}
\end{aligned}
$$

using the Cauchy–Schwarz inequality.

Define $V : \mathscr{H}(K) \to l^2(\mathbb{N})$ by

$$V(G) = \left(\langle G, e_n \rangle_{\mathscr{H}(K)} \right)_{n \in \mathbb{N}}.$$

It is easy to show that V is isometric. Moreover, the adjoint $V^* : l^2(\mathbb{N}) \to \mathscr{H}(K)$ is given by

$$V^*((a_n))(t) = \sum a_n e_n(t) \in \mathscr{H}(K),$$

so that $V^*V = I_{\mathscr{H}(K)}$. Note that (4.19) is a canonical representation, but it is not unique unless (e_n) is assumed to be an ONB and not just a Parseval frame.

More details: Suppose l is a linear functional on $\mathscr{H}(K)$ and $(l(e_n)) \in l^2$, then

$$F = F_l = \sum l(e_n) e_n \in \mathscr{H}(K)$$

is well defined, since $\{e_n\}$ is a Parseval frame. We also have

$$l(G) = \sum \langle G, e_n \rangle_{\mathscr{H}(K)} l(e_n)$$

$$= \left\langle \sum l(e_n) e_n, G \right\rangle_{\mathscr{H}(K)} = \langle F_l, G \rangle_{\mathscr{H}(K)},$$

which is the desired conclusion, i.e., $T_K(l) = F_l$ and $T_K^*(F_l) = l$. □

Example 4.19. Let $S = [0, 1]$, $K(s, t) = s \wedge t$. This is a *relative kernel*, since we specify

$$\mathscr{H}(K) = \{F : F' \in L^2(0, 1), F(0) = 0\},$$

with $\|F\|_{\mathscr{H}(K)}^2 = \int_0^1 |F'(s)|^2 ds$, and we then get

$$\langle F(\cdot), \cdot \wedge t \rangle_{\mathscr{H}(K)} = \int_0^1 F'(s) \chi_{[0,t]}(s) ds$$

$$= \int_0^t F'(s) ds$$

$$= F(t) - F(0) = F(t).$$

In this example,

$$(T_K \mu)(t) = \int \mu(ds) s \wedge t = t - \frac{1}{2} t^2,$$

$$\mu K \mu = \|T_K \mu\|_{\mathscr{H}(K)}^2 = \int_0^1 |(T_K \mu)'(t)|^2 dt = \frac{1}{3},$$

where $(T_k \mu)'(t) = \mu([t, 1])$. See Section 8.2 for details on relative kernels.

Example 4.20. Let $K(s, t) = \frac{1}{1-st}$, $s, t \in (-1, 1)$, and let $\mathscr{H}(K)$ be the corresponding RKHS. The kernel K is p.d. since

$$K(s, t) = \sum_0^\infty s^n t^n = \sum_0^\infty e_n(s) e_n(t),$$

where $e_n(t) = t^n$, and $\{e_n\}$ is an ONB in $\mathscr{H}(K)$. Now fix $n > 0$ and consider the linear functional

$$l(G) = \langle G, e_n(\cdot)\rangle_{\mathscr{H}(K)}. \tag{4.20}$$

Then $|l(G)| \leq \|G\|_{\mathscr{H}(K)}\|e_n\|_{\mathscr{H}(K)} \leq \|G\|_{\mathscr{H}(K)}$, by Schwarz and the fact that $\|e_n\|_{\mathscr{H}(K)} = 1$. It follows that $C_l = 1$.

However, the LHS of (4.20) cannot be realized by a measure on $(-1, 1)$. In fact, the unique solution is given by $l = \frac{1}{n!}\delta_0^{(n)} \in \mathscr{L}_2(K)$, where $\delta_0^{(n)}$ is the Schwartz distribution (not a measure of course): $\delta_0^{(n)}(\varphi) = (-1)^n \varphi^{(n)}(0)$, $\varphi^{(n)} = \left(\frac{d}{dt}\right)^n \varphi$, for all $\varphi \in C_c(-1, 1)$.

In this example, $\delta_0^{(n)} \in \mathscr{L}_2(K)$, but *not* in $\mathfrak{M}_2(K)$. However, it is in the closure of $\mathfrak{M}_2(K)$. The plan is therefore to extend from the space of $\mu = \sum_i c_i\delta_{s_i}$ to the more general measures, and (in some examples) to the case of Schwartz distributions (see, e.g., Section 4.1.2).

Moreover, we have

$$\delta_{x_0} = \sum_{n\in\mathbb{N}_0} \frac{x_0^n}{n!}\delta_0^{(n)}, \quad |x_0| < 1. \tag{4.21}$$

One may apply T_K to both sides of (4.21), and get

$$\text{LHS}_{(4.21)} = T_K(\delta_{x_0})(t) = \frac{1}{1 - x_0 t} = K(x_0, t),$$

$$\text{RHS}_{(4.21)} = \sum_{n\in\mathbb{N}_0} \frac{x_0^n}{n!}\left(T_K\left(\delta_0^{(n)}\right)\right)(t) = \sum_{n\in\mathbb{N}_0} \frac{x_0^n}{n!}n!e_n(t)$$

$$= \sum_{n\in\mathbb{N}_0} x_0^n t^n = K(x_0, t).$$

Note that (4.21) lives in $\mathscr{L}_2(K)$ and so depends on K. We will later consider the case $K(s, t) = e^{st}$.

For more details, also see Example 4.24.

4.1.2 *Function spaces and Schwartz distributions*

> *"What is important is to deeply understand things and their relations to each other. This is where intelligence lies. The fact of being quick or slow isn't really relevant."* — Laurent Schwartz.

This section gives an overview in the case of functions on $\mathbb{R}$, while the statements generalize easily to $\mathbb{R}^n$, and to the case of functions defined on open subsets $\Omega \subset \mathbb{R}^n$.

We shall consider the inclusions:

$$\mathscr{D} \hookrightarrow \mathscr{S} \hookrightarrow \mathscr{E} \tag{4.22}$$

where $\mathscr{D} = C_c^\infty(\mathbb{R})$, $\mathscr{S} = \mathscr{S}(\mathbb{R}^n)$ and $\mathscr{E} = C^\infty(\mathbb{R})$, using Schwartz' terminologies for the three spaces. (See, e.g., [Trè67, Trè06, Sch94, Sch68, Sch66]) Each space is equipped with a locally convex topology defined by the following respective seminorms.

- $\mathscr{D}$: the system of seminorms are indexed by $n \in \mathbb{N}$ and compact subsets K of $\mathbb{R}$, specified by

$$p_{n,K}(\varphi) = \max_{0 \le j \le n} \sup_{t \in K} \left| \varphi^{(j)}(t) \right|; \tag{4.23}$$

- $\mathscr{S}$: the seminorms are indexed by $n, m \in \mathbb{N}_0$, defined as

$$p_{n,m}(\varphi) = \sup_{t \in \mathbb{R}} \left(1 + |t|^n\right) \left| \varphi^{(m)}(t) \right|; \tag{4.24}$$

- $\mathscr{E}$: same seminorms as in (4.23), but the topology is different. $\mathscr{D}$ is given the *inductive limit topology*, while $\mathscr{E}$ is given the *projective limit topology*, indexed by the seminorms from (4.23).

In all three cases for the inclusions (4.22), $\mathscr{D}$ is dense in $\mathscr{S}$, and $\mathscr{S}$ is dense in $\mathscr{E}$. When we pass to the respective dual spaces (of distributions), we get the reversal inclusions:

$$\mathscr{E}' \hookrightarrow \mathscr{S}' \hookrightarrow \mathscr{D}', \tag{4.25}$$

where $\mathscr{E}'$ contains compactly supported distributions, $\mathscr{S}'$ denotes the tempered distributions, and $\mathscr{D}'$ all the Schwartz distributions (See Schwartz, Trves: [Sch66, Trè67, Trè06].)

For $u \in \mathscr{D}'$, we consider the distributional derivative $u^{(n)} := \left(\frac{d}{dt}\right)^n u$ defined as follows, for $\varphi \in \mathscr{D}$:

$$u^{(n)}(\varphi) := (-1)^n u\left(\varphi^{(n)}\right) = (-1)^n u\left(\left(\frac{d}{dt}\right)^n \varphi\right). \tag{4.26}$$

Note the following:

$$u \in \mathscr{E}' \implies u^{(n)} \in \mathscr{E}', \tag{4.27}$$

$$v \in \mathscr{S}' \implies v^{(n)} \in \mathscr{S}', \tag{4.28}$$

$$w \in \mathscr{D}' \implies w^{(n)} \in \mathscr{D}'. \tag{4.29}$$

The standard *Fourier transform* will play a role below both in the context of function spaces and for spaces of distributions.

For example, if $f \in L^2(\mathbb{R})$, set

$$\widehat{f}(t) := \int_{\mathbb{R}} e^{-itx} f(x) \, dx. \tag{4.30}$$

Note that then

$$\frac{1}{2\pi} \int_{\mathbb{R}} \left| \widehat{f}(t) \right|^2 dt = \int_{\mathbb{R}} |f(x)|^2 \, dx = \|f\|_{L^2(\mathbb{R})}^2 . \tag{4.31}$$

Further note that (4.30) extends to the space of Schwartz distributions as follows; for example, if $u \in \mathscr{E}'$ (a distribution of compact support), let $\widehat{u}$ be defined by

$$\widehat{u}(t) := u \, (\text{acting in } x) \left(e^{-itx} \right). \tag{4.32}$$

More generally, if $u \in \mathscr{D}'$, and $\varphi \in \mathscr{D}$, set

$$\widehat{u}(\varphi) := u \left(\varphi^{\vee} \right),$$

where

$$\varphi^{\vee}(t) = \int_{\mathbb{R}} e^{itx} \varphi(x) \, dx.$$

Extending (4.27)–(4.29), we have the implication:

$$u \in \mathscr{S} \implies \widehat{u} \in \mathscr{S}'. \tag{4.33}$$

For $\varphi, \psi \in \mathscr{S}$, set

$$\psi(\varphi) := \int_{\mathbb{R}} \psi(x) \, \varphi(x) \, dx.$$

With this action, note that $\mathscr{S}$ is then identified with a subspace of $\mathscr{S}'$ (the tempered distributions). One checks that with respect to the dual topology on $\mathscr{S}'$, we get that $\mathscr{S}$ is in fact dense.

Moreover, with these implications, we further have

$$\mathscr{S} \hookrightarrow L^2(\mathbb{R}) \hookrightarrow \mathscr{S}'; \tag{4.34}$$

again with each inclusion in (4.34) referring to "dense subspace." In fact, using that $\mathscr{S}$ is dense in $L^2(\mathbb{R})$ (with respect to Hilbert norm in $L^2(\mathbb{R})$), we conclude that the dual "inclusion" $L^2(\mathbb{R}) \hookrightarrow \mathscr{S}'$ is an injection, so 1-1.

The triple inclusion (4.34) is called a *Gelfand-triple*. It is important that $\mathscr{S} \hookrightarrow L^2(\mathbb{R})$ is a nuclear mapping.

Definition 4.21. Let V be a vector space. A function N on V is said to be *conditionally negative definite* (CND) if, for all finite systems $(c_i)_{i=1}^n$,

$(v_i)_{i=1}^n$, $c_i \in \mathbb{R}$, $\sum c_i = 0$, $v_i \in V$, we have

$$\sum_i \sum_j c_i c_j N(v_i - v_j) \leq 0. \tag{4.35}$$

Theorem 4.22 (Schoenberg. See, e.g., [AESW51, Sch42, vNS41, Sch37, Sch38]). *A function $N : V \to \mathbb{R}$ is conditionally negative definite (CND) if there is a Hilbert space $\mathscr{H}$ with norm $\|\cdot\|_{\mathscr{H}}$, and a linear function, $\psi : V \to \mathscr{H}$, such that*

$$N(u,v) = \|\psi(u) - \psi(v)\|_{\mathscr{H}}^2, \ v \in V.$$

We now turn to the Gelfand-triple from (4.34).

Theorem 4.23 (Bochner–Minlos. See, e.g., [Ok08, OR00, Ban97]). *Let $N : \mathscr{S} \to \mathbb{R}$ be a continuous (CND) function. Then there is a Gaussian process $\{X_\varphi\}_{\varphi \in \mathscr{S}}$, realized in the probability space $(\mathscr{S}', \mathscr{F}_{cyl}, \mathbb{P})$, where $\mathscr{F}_{cyl}$ denotes the cylinder σ-algebra, such that $\mathbb{E}(X_\varphi^{(N)}) = 0$ for all $\varphi \in \mathscr{S}$, where $\mathbb{E}(\cdots) = \int_{\mathscr{S}'} (\cdots) \, d\mathbb{P}$, and*

$$\mathbb{E}\left(e^{iX_\varphi}\right) = e^{-\frac{1}{2}N(\varphi)}, \ \varphi \in \mathscr{S}.$$

In particular, we have

$$\mathbb{E}\left(\left(X_\varphi^{(N)}\right)^2\right) = N(\varphi), \ \varphi \in \mathscr{S}.$$

Example 4.24. Let $K : (-1,1) \times (-1,1) \to \mathbb{R}$ (see Example 4.20),

$$K(s,t) = \frac{1}{1-st} = \sum_{n=0}^\infty s^n t^n.$$

Set $e_n(t) = t^n$. Then $\{e_n\}_{n \in \mathbb{N}_0}$ is an ONB in the RKHS $\mathscr{H}(K)$. Hence, functions in $\mathscr{H}(K)$ have unique representations

$$F(s) = \sum_{n=0}^\infty \alpha_n s^n, \ (\alpha_n) \in l^2(\mathbb{N}_0), \ s \in (-1,1).$$

For a compactly supported Schwartz distribution l in $(-1,1)$, we shall consider

$$T_K(l) := l_{(s)}(K(s,t)),$$

i.e., l acts in the first variable. By direct computation, one has

$$(-1)^n \left(\frac{d}{ds}\right)^n K(s,\cdot) = \frac{(-1)^n n! t^n}{(1-st)^{n+1}}.$$

Evaluation at $s = 0$ yields $(-1)^n n! t^n$. Thus the distribution derivative applied to the Dirac measure δ_0 gives

$$\delta_0^{(n)}(s)\left(K(s,t)\right) = (-1)^n n! t^n,$$

and so we get

$$T_K\left(\delta_0^{(n)}\right)(t) = (-1)^n n! e_n(t).$$

4.1.3   The isomorphism $T_K : \mathscr{L}_2(K) \to \mathscr{H}(K)$

Motivation. The RKHS $\mathscr{H}(K)$ is defined abstractly wheres $\mathfrak{M}_2(K)$ and $\mathscr{L}_2(K)$ are explicit (see Def. 4.15 & 4.16), and more "natural". Starting with a p.d. kernel K on S, then the goal is to identify Hilbert spaces of measures $\mathfrak{M}_2(K)$, or Hilbert spaces of Schwartz distributions, say $\mathscr{L}_2(K)$, which can give a more explicit norm, with a natural isometry $\mathscr{L}_2(K) \xrightarrow{T_K} \mathscr{H}(K)$.

The mapping T_K may be defined for measures μ, viewed as elements in $\mathscr{L}_2(K)$, as the generalized integral operator:

Definition 4.25. For $\mu \in \mathfrak{M}_2(K)$, set

$$(T_K\mu)(t) = \int \mu(ds)\,K(s,t), \quad \mu \in \mathfrak{M}_2(K). \tag{4.36}$$

For $l \in \mathscr{L}_2(K)$, by Riesz (Lemma 1.18) there exists a unique $F = F_l \in \mathscr{H}(K)$ such that $l(G) = \langle F_l, G\rangle_{\mathscr{H}(K)}$, for all $G \in \mathscr{H}(K)$.

Definition 4.26. For $l \in \mathscr{L}_2(K)$, set

$$T_K(l) = F_l, \quad \text{and so} \quad T_K^*(F_l) = l. \tag{4.37}$$

Alternatively, one may define

$$F_l := T_K(l) = l\,(\text{acting in } s)\,(K(s,\cdot)), \tag{4.38}$$

by the reproducing property.

Lemma 4.27. *Let K be p.d. on $S \times S$ with the cylinder σ-algebra $\mathscr{B}_{(S,K)}$. Let μ be a positive measure on $\mathscr{B}_{(S,K)}$ such that $\int K(s,t)\mu(ds) < \infty$, then $L^2(\mu) \hookrightarrow \mathscr{H}(K)$ is a bounded embedding, i.e., $f d\mu \in \mathfrak{M}_2(K)$, for all $f \in L^2(\mu)$.*

Proof. Given $f \in L^2(\mu)$, then

$$\left\| \int f(s) K(s,\cdot) \mu(ds) \right\|_{\mathscr{H}(K)} \leq \int |f(s)| \, \|K(s,\cdot)\|_{\mathscr{H}(K)} \, \mu(ds)$$

$$\overset{\text{(Schwarz)}}{\leq} \|f\|_{L^2(\mu)} \left(\int K(s,s) \mu(ds) \right)^{1/2}$$

$$= C_\mu^{1/2} \|f\|_{L^2(\mu)} \, .$$

Hence, $\int K(s,\cdot) f(s) \mu(ds) = T_K(f d\mu) \in \mathscr{H}(K)$ and

$$\|T_K(f d\mu)\|_{\mathscr{H}(K)}^2 = \iint f(s) K(s,t) f(t) \mu(ds) \mu(dt)$$

$$= (f d\mu) K (f d\mu)$$

$$= \|f d\mu\|_{\mathfrak{M}_2(K)}^2 \, . \qquad \square$$

Lemma 4.28. *If $\mu \in \mathfrak{M}_2(K)$, then*

$$(T_K \mu)(t) = \int \mu(ds) K(s,t)$$

is well defined, and $(T_K \mu)(\cdot) \in \mathscr{H}(K)$. Moreover, $T_K : \mathfrak{M}_2(K) \to \mathscr{H}(K)$ is isometric, where

$$\|T_K \mu\|_{\mathscr{H}(K)}^2 = \|\mu\|_{\mathfrak{M}_2(K)}^2 = \iint \mu(ds) K(s,t) \mu(dt). \tag{4.39}$$

Proof. With the assumption $\mu \in \mathfrak{M}_2(K)$, the exchange of $\langle \cdot, \cdot \rangle_{\mathscr{H}(K)}$ and integral with respect to μ can be justified. Thus,

$$\text{LHS}_{(4.39)} = \langle T_K \mu, T_K \mu \rangle_{\mathscr{H}(K)}$$

$$= \int (T_K \mu)(t) \mu(dt)$$

$$= \iint \mu(ds) K(s,t) \mu(dt)$$

$$= \mu K \mu = \|\mu\|_{\mathfrak{M}_2(K)}^2 = \text{RHS}_{(4.39)}. \qquad \square$$

Definition 4.29. By a partition $\{E_i\}$ of S, we mean $E_i \in \mathscr{B}_{(S,K)}$ of finite measure, $E_i \cap E_j = \emptyset$ if $i \neq j$, and $\cup E_i = S$.

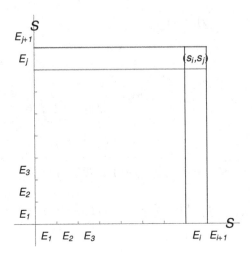

Figure 4.1: $\mathscr{P}$: partitions of S, $E_i \in \mathscr{B}_{(S,K)}$.

Lemma 4.30. *Let μ be a signed measure, then $\mu \in \mathfrak{M}_2(K)$ if and only if*

$$\left\| \sum_i \mu(E_i) K(s_i, \cdot) \right\|_{\mathscr{H}(K)}^2 = \sup_{\text{all partitions}} \sum_i \sum_j \mu(E_i) K(s_i, s_j) \mu(E_j) < \infty$$

with $s_i \in E_i$. In this case, we have the following norm limit

$$\lim \left\| T_K \mu - \sum_i \mu(E_i) K(s_i, \cdot) \right\|_{\mathscr{H}(K)} = 0,$$

where the limit is taken over filters of partitions. See Figure 4.1.

Proof. For $\mu \in \mathfrak{M}_2(K)$, we must make precise $\int G(s)\mu(ds)$ as a well defined integral. Again, we take limits over all partitions $\mathscr{P}$ of S:

$$\sum_i G(s_i)\mu(E_i) = \sum \mu(E_i) \langle K(s_i, \cdot), G(\cdot) \rangle_{\mathscr{H}(K)}$$

$$= \left\langle \sum \mu(E_i) K(s_i, \cdot), G(\cdot) \right\rangle_{\mathscr{H}(K)} \to \langle T_K \mu, G \rangle_{\mathscr{H}(K)},$$

and so $T_K \mu$ is the $\|\cdot\|_{\mathscr{H}(K)}$-norm limit of $\sum \mu(E_i) K(s_i, \cdot)$.

More specifically, given a partition $P = \{E_i\} \in \mathscr{P}$, let

$$T(P) := \sum_i \mu(E_i) K(s_i, \cdot) \in \mathscr{H}(K),$$

then we have

$$\|T(P)\|^2_{\mathscr{H}(K)} = \sum_i \sum_j \mu(E_i) K(s_i, s_j) \mu(E_j)$$

and $T(P) \to T_K(\mu)$, as a filter in partitions. $\square$

Lemma 4.31. *For all $G \in \mathscr{H}(K)$ and $\mu \in \mathfrak{M}_2(K)$, we have:*

$$\int G(t)\,\mu(dt) = \langle G, T_K\mu \rangle_{\mathscr{H}(K)}. \qquad (4.40)$$

Proof.

$$\begin{aligned}
\text{LHS}_{(4.40)} &= \int \langle G(\cdot), K(\cdot, t) \rangle_{\mathscr{H}(K)}\,\mu(dt) \\
&= \left\langle G, \int \mu(dt)\, K(\cdot, t) \right\rangle_{\mathscr{H}(K)} \\
&= \langle G, T_K\mu \rangle_{\mathscr{H}(K)}.
\end{aligned}$$

We conclude that $\int G(t)\,\mu(dt) = \langle T_K\mu, G \rangle_{\mathscr{H}(K)} \in \mathscr{H}(K)$. Note that since $\mu \in \mathfrak{M}_2(K)$, we may exchange integral with $\langle \cdot, \cdot \rangle_{\mathscr{H}(K)}$. $\square$

Lemma 4.32. *Let $\left(S, \mathscr{B}_{(S,K)}, K\right)$ be as above, and consider signed measures μ on $\left(S, \mathscr{B}_{(S,K)}\right)$. Then*

$$\mu \in \mathfrak{M}_2(K) \Longleftrightarrow \left| \int G d\mu \right| \leq C_\mu \|G\|_{\mathscr{H}(K)}, \ \forall G \in \mathscr{H}(K).$$

Proof. Given $\mu \in \mathfrak{M}_2(K)$, then

$$\begin{aligned}
\left| \int G(t)\,\mu(dt) \right| &= \left| \int \langle K(\cdot, t), G(\cdot) \rangle_{\mathscr{H}(K)}\,\mu(dt) \right| \\
&\leq \left| \left\langle \int K(\cdot, t)\,\mu(dt), G \right\rangle_{\mathscr{H}(K)} \right| \\
&= \left| \langle T_K\mu, G \rangle_{\mathscr{H}(K)} \right| \\
&\leq \|T_K\mu\|_{\mathscr{H}(K)} \|G\|_{\mathscr{H}(K)},
\end{aligned}$$

and so $C_\mu = \|T_K\mu\|_{\mathscr{H}(K)} < \infty$.

Conversely, suppose $\left| \int G d\mu \right| \leq C_\mu \|G\|_{\mathscr{H}(K)}$ for all $G \in \mathscr{H}(K)$. Then by applying Lemma 1.18 to $\mathscr{H}(K)$, there exists a unique $F \in \mathscr{H}(K)$ such that

$$\int G d\mu = \langle G, F \rangle_{\mathscr{H}(K)}, \quad \forall G \in \mathscr{H}(K).$$

Now take $G = K(\cdot, t)$, and by the reproducing property, we get

$$\underbrace{\int \mu(ds) K(s,t)}_{T_K \mu} = \langle K(\cdot, t), F \rangle_{\mathscr{H}(K)} = F(t),$$

where $T_K \mu = F \in \mathscr{H}(K)$ by Riesz (Lemma 1.18), and $\mu = T_K^* F$. We conclude that $\mu \in \mathfrak{M}_2(K)$.   □

In summary, the goal is to pass from (1) atomic measures to (2) σ-finite signed measures, and then to (3) linear functionals on $\mathscr{H}(K)$, continues with respect to the $\|\cdot\|_{\mathscr{H}(K)}$ norm. The respective mappings are specified as follows:

$$T_K : \sum_i c_i \delta_i \longmapsto \sum_i c_i K(s_i, \cdot) \in \mathscr{H}(K); \tag{4.41}$$

$$(T_K \mu)(t) = \int \mu(ds) K(s, \cdot) \in \mathscr{H}(K), \quad \mu \in \mathfrak{M}_2(K); \tag{4.42}$$

$$(T_K l)(t) = l.(K(\cdot, t)) \in \mathscr{H}(K), \quad l \in \mathscr{L}_2(K). \tag{4.43}$$

In all cases (l as atomic measures, signed measures, or the general case of linear functionals), the mapping $T_K : \mathscr{L}_2(K) \to \mathscr{H}(K)$ is an isometry, i.e., $\|T_K l\|_{\mathscr{H}(K)} = \|l\|_{\mathscr{L}_2(K)}$. In general, we have

$$T_K(\mathfrak{M}_2(K)) \subseteq T_K(\mathscr{L}_2(K)) = \mathscr{H}(K), \tag{4.44}$$

where $T_K(\mathfrak{M}_2(K))$ is dense in $\mathscr{H}(K)$ in the $\|\cdot\|_{\mathscr{H}(K)}$ norm. In some cases,

$$T_K(\mathfrak{M}_2(K)) = \mathscr{H}(K); \tag{4.45}$$

but not always.

Remark 4.33. $\mathfrak{M}_2(K)$ and factorizations on K specified by a positive measure M on X. Let

$$K(s,t) = \int \psi(s, x) \psi(t, x) M(dx),$$

where $\psi(s, \cdot) \in L^2(M)$. Let μ be a signed measure on S. Then

$$\mu \in \mathfrak{M}_2(K) \Longleftrightarrow T_\mu \psi \in L^2(M)$$

where $(T_\mu \psi)(x) = \int \mu(ds) \psi(s,x)$. Indeed, using Fubini, one has

$$\iint \mu(ds) K(s,t) \mu(dt) = \int \left(\int \mu(ds) \psi(s,x) \int \mu(dt) \psi(t,x) \right) M(dx)$$

$$= \int |(T_\mu \psi)(x)|^2 M(dx)$$

$$= \|T_K(\mu)\|_{\mathcal{H}(K)}^2.$$

Example 4.34 ($\mathfrak{M}_2(K)$ vs $\mathcal{L}_2(K)$). For $K(s,t) = s \wedge t$ on $[0,1]$, Equation (4.45) holds, but not for $K(s,t) = \frac{1}{1-st}$, $s,t \in (-1,1)$.

Here, with $K(s,t) = s \wedge t$ and $d\mu = F(t)\,dt$, one has $\int_0^1 G(t)\,d\mu(t) = \langle G, T_K \mu \rangle_{\mathcal{H}(K)}$, for all $G \in \mathcal{H}(K)$. Set $G = F$, then

$$\int_0^1 F\,d\mu = \langle F, F \rangle_{\mathcal{H}(K)}$$

$$= \int_0^1 \int_0^1 F(s)(s \wedge t) F(t)\,ds\,dt = \int_0^1 |f(t)|^2\,dt,$$

where $f = F'$ is the distributional derivative of F. It follows that $\mu \ll \lambda$, the usual Lebesgue measure.

Lemma 4.35. *If K is p.d. on S and if the induced metric $d_K(s,t) = \|K_s - K_t\|_{\mathcal{H}(K)}$ is bounded, then $\mathfrak{M}_2(K)$ is complete, and so $T_K(\mathfrak{M}_2(K)) = \mathcal{H}(K)$.*

Proof. See [JT18a] for details. Sketch: With the metric d_K, we can complete, and then use Stone–Weierstrass. Indeed, if l is continuous in the $\|\cdot\|_{\mathcal{H}(K)}$ norm, then there exists a signed measure μ such that $l(G) = \int G(t)\,\mu(dt)$. $\qquad \square$

An overview of the relation between $\mathfrak{M}_2(K)$ and $\mathcal{L}_2(K)$.

Let $K : S \times S \to \mathbb{R}$ be a given p.d. kernel on S, and consider the RKHS $\mathcal{H}(K)$, as well as the isometry T_K introduced above. We have:

$$T_K(\mathfrak{M}_2(K)) \underset{\text{possible} \neq}{\subseteq} T_K(\mathcal{L}_2(K)) = \mathcal{H}(K)$$

and $T_K(\mathfrak{M}_2(K))$ is dense in $\mathcal{H}(K)$ in the $\|\cdot\|_{\mathcal{H}(K)}$ norm.

By Example 4.34, we see that $T_K(\mathfrak{M}_2(K))$ may not be closed in $\mathcal{H}(K)$, and so $\mathfrak{M}_2(K)$ may not be complete. However, by the very construction,

$\mathscr{L}_2(K)$ will be complete, and $T_K(\mathscr{L}_2(K)) = \mathscr{H}(K)$. See Lemma 4.36 and Lemma 4.37.

Lemma 4.36. $T_K(\mathscr{L}_2(K))$ *is complete.*

Proof. Suppose $\{\mu_n\}$ is Cauchy in $\mathscr{L}_2(K)$, $\|\mu_n - \mu_m\|_{\mathscr{L}_2(K)} \to 0$, then $F_n = T_K \mu_n$ satisfy $\|F_n - F_m\|_{\mathscr{H}(K)} \to 0$, and so there exists $F \in \mathscr{H}(K)$ such that $\|F_n - F\|_{\mathscr{H}(K)} \to 0$. Define $l(G) := \langle F, G \rangle_{\mathscr{H}(K)}$, for all $G \in \mathscr{H}(K)$. Then, $l \in \mathscr{L}_2(K)$ and $\|\mu_n - l\|_{\mathscr{L}_2(K)} \to 0$. $\square$

Lemma 4.37. *The isometry* T_K *maps* $\mathscr{L}_2(K)$ *onto* $\mathscr{H}(K)$.

Proof. We shall show that $ker(T_K^*) = 0$. Suppose $T_K^*(G) = 0$. Then for all μ,

$$0 = \langle \mu, T_K^* G \rangle_{\mathfrak{M}_2(K)} = \langle T_K \mu, G \rangle_{\mathscr{H}(K)}.$$

Now take $\mu = \delta_t$, $t \in S$; then

$$0 = \langle T_K(\delta_t), G \rangle_{\mathscr{H}(K)} = \langle K(\cdot, t), G(\cdot) \rangle_{\mathscr{H}(K)} = G(t).$$

Thus, $G(t) = 0$ for all $t \in S$, and so $G = 0$ in $\mathscr{H}(K)$. $\square$

Question 4.38. If $\Omega \subset \mathbb{R}^d$ is open and $K : \Omega \times \Omega \to \mathbb{R}$ is positive definite, when is $\mathscr{H}(K)$ locally (or globally) translation invariant?

Example 4.39.

(i) For $K(s,t) = \frac{1}{1-st}$, $s,t \in (-1,1)$, the RKHS $\mathscr{H}(K)$ is locally translation invariant.

(ii) The case of stationary kernels $K(s,t) = K(s-t)$ is especially important, since $\mathbb{R}^d$-translation invariance yields a unitary operator, with $\mathscr{H}(K) = $ distribution of f, such that $\|f\|_{\mathscr{H}(K)}^2 = \int |\widehat{f}(u)|^2 M(du)$, where M is a Borel measure such that

$$K(s-t) = \int e^{i(s-t)u} M(du),$$

and $\widehat{U_t f}(v) = e^{itv} \widehat{f}(v)$.

(iii) $K(s,t) = e^{s \cdot t}$

4.1.4 *Reversible kernels*

Since, in general, a reproducing kernel Hilbert space (RKHS), defined from a given positive definite (p.d.) kernel K, is a rather abstract construction, one seeks special cases where explicit transforms are available. For this endeavor, the object is to identify p.d. kernels K for which the associated RKHS $\mathscr{H}(K)$ admits an explicit transform; so a transform for the RKHS (K) will then be realized explicitly in a suitable $L^2(\lambda)$.

For this, we consider the special case of a transition kernel K (measures in the second variable) which has a measure λ, which turns K into a reversible Markov transition. When this happens, we shall use the notation RKHS (K, λ); or just $\mathscr{H}(K, \lambda)$. Then there is a λ Radon–Nikodym derivative of K in the second variable, and we get a corresponding Radon–Nikodym derivative function, say F. It is realized as a p.d. function F which is measurable, now as a function in two variables.

To appreciate the condition on a given pair (K, λ), note that we have equivalence of the following two assertions (i) and (ii) where: (i) a transition kernel K has a measure λ, which turns it into a reversible transition. And (ii): the kernel K induces a symmetric operator in $L^2(\lambda)$ (the operator is also denoted K).

In general, this induced operator K is unbounded but with dense domain in $L^2(\lambda)$. If the operator K in $L^2(\lambda)$ is positive semi-definite, then it has a Friedrichs extension $\widetilde{K}$ (see Remark 1.71, and [DS88]). The operator $\widetilde{K}$ is a semibounded selfadjoint extension of K. We shall use the square root of $\widetilde{K}$ in order to define an explicit transform for the RKHS(K, λ).

The presentation of this general setting is followed by a discussion of some examples/special cases, Mercer operators, and the generalized Brownian motion kernels.

Mercer Kernels

In general for a Mercer operator K acting on $L^2(\lambda)$, we mean

$$Kf(t) = \int K(t, ds) f(s), \ f \in L^2(\lambda),$$

where $K(t, \cdot)$ is a measure for every t.

Assume $K(t, \cdot) \ll \lambda$ with Radon–Nikodym derivative

$$\frac{K(t, \cdot)}{d\lambda}(s) = F(t, s),$$

then

$$Kf(t) = \int F(t,s) f(s) \, ds.$$

Further assume that K is compact and positive (selfadjoint), and let

$$K(s,t) = \sum \lambda_n \varphi_n(s) \varphi_n(t) \tag{4.46}$$

be its spectral decomposition, i.e., $K\varphi_n = \lambda_n \varphi_n$, and $\{\varphi_n\}$ is an ONB in $L^2(\lambda)$. Then

$$\|f\|^2_{\mathscr{H}(K)} = \sum \lambda_n \left| \langle \varphi_n, f \rangle_{L^2(\lambda)} \right|^2$$

$$= \langle f, Kf \rangle_{L^2(\lambda)} = \left\| K^{1/2} f \right\|^2_{L^2(\lambda)}.$$

Lemma 4.40 (Karhunen–Loève). *Let K be as in (4.46). Let $\{Z_n\}$ be i.i.d. random variables with $Z_n \sim N(0,1)$. Set $X_t = \sum \sqrt{\lambda_n} \varphi_n(t) Z_n(\cdot)$; then*

$$\mathbb{E}(X_s X_t) = K(s,t).$$

Proof. We have

$$\mathbb{E}(X_s X_t) = \sum_m \sum_n \sqrt{\lambda_m} \sqrt{\lambda_n} \varphi_m(s) \varphi_n(t) \mathbb{E}(Z_m Z_n)$$

$$= \sum_n \lambda_n \varphi_n(s) \varphi_n(t) \mathbb{E}(Z_n^2)$$

$$= \sum_n \lambda_n \varphi_n(s) \varphi_n(t) = K(s,t).$$

$\square$

Example 4.41. The standard Brownian bridge in $[0, \pi]$ can be represented as

$$X_t = \sqrt{\frac{2}{\pi}} \sum_{n=1}^{\infty} \frac{\sin nt}{n} Z_n(\cdot)$$

where $Z_n(\cdot)$ is i.i.d. and $Z_n \sim N(0,1)$. See Figure 4.2. Then

$$K(s,t) = \mathbb{E}(X_s X_t) = \frac{2}{\pi} \sum_{n=1}^{\infty} \frac{\sin ns}{n} \frac{\sin nt}{n} = s \wedge t - \frac{st}{\pi}. \tag{4.47}$$

Note that $X_0 = X_\pi = 0$, and so X_t is a pinned Brownian motion.

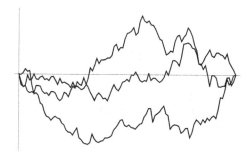

Figure 4.2: Brownian bridge pinned at 0 and π with three sample paths.

If $L = -\left(\frac{d}{dt}\right)^2$, then K satisfies

$$L.K\left(\cdot, t\right) = \delta\left(\cdot - t\right);$$

equivalently, $(LKf)(t) = f(t)$. Hence K is the Greens function for L, and it inverts the elliptic operator L. Also,

$$trace\left(L\right) = \int_0^\pi K\left(t, t\right) dt$$

$$= \frac{2}{\pi} \sum_{n=1}^\infty \frac{1}{n^2} \int_0^\pi \sin^2\left(nt\right) dt$$

$$= \sum_{n=1}^\infty \frac{1}{n^2} = \frac{\pi^2}{6}.$$

Example 4.42. The covariance kernel of standard Brownian motion in $[0, \pi]$ has the representation:

$$K\left(s, t\right) = \frac{2}{\pi} \sum_{n=0}^\infty \frac{\sin\left((n+1/2)\, s\right) \sin\left((n+1/2)\, t\right)}{(n+1/2)^2} = s \wedge t. \qquad (4.48)$$

See Figure 4.3.

Consider $K\left(s, t\right) = s \wedge t$ acting in $L^2(0, b)$, by

$$(Kf)\left(t\right) = \int_0^b t \wedge s f\left(s\right) ds, \quad f \in L^2\left(0, b\right).$$

Since

$$(Kf)\left(t\right) = \int_0^t s f\left(s\right) ds + t \int_t^b f\left(s\right) ds,$$

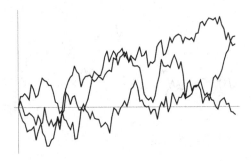

Figure 4.3: Standard Brownian motion in $[0, \pi]$ with three sample paths ($\sigma = 0.5$).

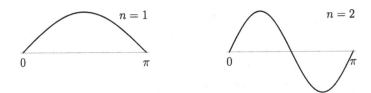

Figure 4.4: Sample eigenfunctions of L.

we get that

$$\frac{d}{dt}(Kf)(t) = \int_t^b f(s)\,ds, \quad -\left(\frac{d}{ds}\right)^2 (Kf)(t) = f(t).$$

So if $Kf = \lambda f$, then

$$-f = \lambda f''.$$

If $f(t) = \cos(pt)$ or $f(t) = \sin(pt)$ (see Figure 4.4), then $p^2 = \lambda^{-1}$. In the special case when $b = \pi$, we get (4.48).

Let $K(s,t) = s \wedge t$, $s, t \in [0,1]$, which is p.d. Let $\mathscr{H}(K)$ be the associated RKHS. Indeed,

$$\mathscr{H}(K) = \{F : F' \in L^2\}.$$

Let $\mathfrak{M}_2(K)$ be as in Definition 4.16.

From the above discussion, we see that, for every $x \in (0,1)$, the Dirac measure δ_x is in $\mathfrak{M}_2(K)$, and $F := T_K \delta_x = K(x, \cdot) = x \wedge \cdot$ on $[0,1]$, with $f := F' = \chi_{[0,x]}(\cdot)$. This is summarized in Table 4.1.

Lemma 4.43. *For the p.d. kernel $K(s,t) = s \wedge t$ on $[0,1]$, we have*

$$\mu \in \mathfrak{M}_2(K) \iff \mu([t,1]) \in L^2(0,1).$$

Table 4.1: The positive kernel for Brownian motion, and its RKHS analysis.

$K(s,t) = s \wedge t$		
$\mu \in \mathfrak{M}_2(K)$ δ_x	$F \in \mathscr{H}(K)$ $= \{F : F' \in L^2\}$ $T_K\delta_x = K(x,\cdot) = x \wedge \cdot$ on $[0,1]$	$f = F' \in L^2$ $f = \chi_{[0,1]}(\cdot)$
	The general case	
$\mu \in \mathfrak{M}_2(K)$ $\mu(ds) = f(s)\,ds$	$F = (T_K\mu)$ $= \int_0^t s\mu(ds) + t\mu([t,1])$ $F = \int_0^t sf(s)\,ds$ $+ t\int_t^1 f(s)\,ds$	$f(t) = \mu([t,1])$ $= (T_K\mu)'(t)$ $F'(t) = f(t) = \int_t^1 f(s)\,ds$

Proof. Note that

$$F = (T_K\mu)(t) = \int_0^t s\mu(ds) + t\int_t^1 d\mu$$

$$= \int_0^t s\mu(ds) + t\mu([t,1])$$

and $F' = \mu([t,1])$. □

Example 4.44. $d\mu = \frac{1}{1-s}ds$ is *not* in $\mathfrak{M}_2(K)$.

The general case

Definition 4.45. Let K be a p.d. kernel on a measure space $(S, \mathscr{B})$. Suppose that $K(x,\cdot)$ is a measure for every $x \in S$, and there exists a σ-finite positive measure λ, such that

$$\lambda(dx)K(x,dy) = \lambda(dy)K(y,dx)$$

$$\Updownarrow \tag{4.49}$$

$$\int_A \lambda(dx)K(x,B) = \int_B \lambda(dy)K(y,A), \quad \forall A,B \in \mathscr{B}.$$

We say that K is a reversible kernel on $(S, \mathscr{B})$, and (K, λ) is a reversible pair.

Remark 4.46. In terms of Markov processes $(\xi_n)_{n\in\mathbb{N}}$, equation (4.49) is equivalent to

$$\text{Prob}(\xi_{n+1} \in B \mid \xi_n \in A) = \text{Prob}(\xi_n \in A \mid \xi_{n+1} \in B). \qquad (4.50)$$

See Figure 4.5 for illustration. In details, there is a path space in $X^\mathbb{N}$ and a path space measure $\mathbb{P}$ defined on (K, λ). For $(x_i) \in X^\mathbb{N}$, there exists a Markov process, $\xi_n((x_i)) = x_n$, such that (4.50) holds.

In the simple case (see Figure 4.6) of finite number of states $n = 1, 2, \ldots, D$, the transition kernel is a double-indexed stochastic matrix $K = (K_{ij})$, with $i, j = 1, \ldots, D$, with $\sum_j K_{ij} = \sum_i K_{ij} = 1$. Then, we have:

$$\text{Prob}(i \to j) = K_{ij}, \quad \text{Prob}(j \to i) = K_{ji},$$

where $K_{ji} = (K^{\text{Tr}})_{ij}$. In this setting, reversibility is equivalent to $K = K^{\text{Tr}}$, i.e., K is symmetric.

Let $\mathscr{B}_{fin} = \{A \in \mathscr{B} : \lambda(A) < \infty\}$. On $\mathscr{D} = span\{\chi_A : A \in \mathscr{B}_{fin}\}$, set

$$(K\varphi)(x) = \int K(x, dy)\,\varphi(y).$$

(If $\varphi = \sum c_i \chi_{A_i}$, a simple function, then $K\varphi = \sum c_i K(x, A_i)$.)

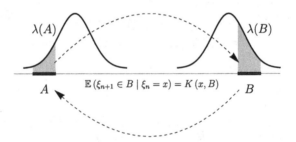

Figure 4.5: Illustration of reversible transitions.

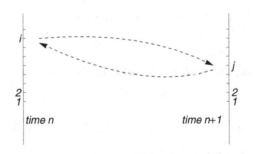

Figure 4.6: A finite reversible Markov process.

Note that $\langle \varphi, K\varphi \rangle_{L^2(\lambda)} \geq 0$, so K is a symmetric operator with dense domain $\mathscr{D}$ in $L^2(\lambda)$, and it has a Fredrichs extension $\widetilde{K}$. The operator $\widetilde{K}$ is selfadjoint and $spec(\widetilde{K}) \subseteq [0, \infty)$.

Set

$$F(A) = \int f(x) K(x, A) \lambda(dx), \quad \text{or} \quad F = \widetilde{f}.$$

By the spectral theorem,

$$\widetilde{K} = \int \lambda Q(d\lambda),$$

where $Q(d\lambda)$ is the corresponding projection valued measure (projection in $L^2(\lambda)$), and

$$(\widetilde{K}f)(t) = \int \lambda(Q(d\lambda)f)(t), \quad f \in L^2(\lambda).$$

For $f, g \in L^2(\lambda)$, we have

$$
\begin{aligned}
\left\langle f, \widetilde{K}g \right\rangle_{L^2(\lambda)} &= \int \overline{f(t)}(\widetilde{K}g)(t)\lambda(dt) \\
&= \int_S \int_S \overline{f(t)} K(t, ds) g(s)\lambda(dt) \\
&= \int_0^\infty \lambda \langle f, Q(d\lambda)g \rangle_{L^2(\lambda)} \\
&= \int_0^\infty \lambda \langle Q(d\lambda)f, Q(d\lambda)g \rangle_{L^2(\lambda)} \\
&= \iint_{S \times S} \int_0^\infty \lambda \overline{(Q(d\lambda)f)(s)}(Q(d\lambda)g)(t)\lambda(ds)\lambda(dt),
\end{aligned}
$$

where

$$K(s, t) = \int_0^\infty \lambda Q(s, d\lambda) Q(t, d\lambda).$$

In the case of measurable kernels, where $K(x, \cdot)$, $x \in X$, is a measure, we then define a function

$$\rho(A \times B) = \int_A \lambda(dx) K(x, B), \quad A, B \in \mathscr{B}_X,$$

and extend ρ to a measure on $X \times X$ with respect to the product σ-algebra.

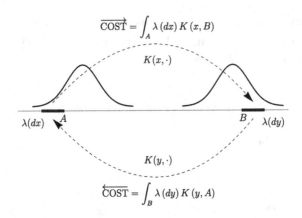

Figure 4.7: Transportation cost.

Then the pair (λ, K) is reversible if and only if $\rho(A \times B) = \rho(B \times A)$, and so

$$\int_A \lambda(dx) K(x, B) = \int \lambda(dy) K(y, A), \quad A, B \in \mathscr{B}_X.$$

The name *transportation cost* is used; see Figure 4.7.

Remark 4.47. In Section 8.2, we will consider electric networks with conductance c_{xy}, $x, y \in V$, the discrete vertex set. Set $c(x) = \sum_{y \sim x} c_{xy}$ and $p_{xy} = c_{xy}/c(x)$, then we get reversible transitions:

$$c(x) p_{xy} = c(y) p_{yx} \quad x \overset{\longrightarrow}{\underset{\longleftarrow}{}} y$$

Remark 4.48. Equation (4.46) corresponds to the case when the operator K has discrete spectrum. Using Dirac's notation, it can be written as

$$K = \sum_1^\infty \lambda_n |\varphi_n\rangle\langle\varphi_n| = \int \lambda Q(d\lambda),$$

with $Q(\{\lambda_n\}) = |\varphi_n\rangle\langle\varphi_n|$ and $Q(\{\lambda_n\}) f = \langle\varphi_n, f\rangle_{L^2(\lambda)} \varphi_n$, for all $f \in L^2(\lambda)$.

Example 4.49. Let $\left\{W_A^{(\lambda)}\right\}_{A \in \mathscr{B}_{fin}}$ be the Wiener process indexed by $\mathscr{B}_{fin}$, which is determined by the covariance

$$K(A, B) := \mathbb{E}\left(W_A^{(\lambda)} W_B^{(\lambda)}\right) = \lambda(A \cap B), \quad A, B \in \mathscr{B}_{fin}.$$

For $f \in L^2(\lambda)$, let $W_f^{(\lambda)} = \int_S f(s) \, dW_s^{(\lambda)}$ defined as an Itô-integral. Then $X_f^{(K,\lambda)} := W_{\tilde{K}^{1/2} f}^{(\lambda)}$ is the Gaussian process corresponding to the (K, λ). Indeed, one has

$$\mathbb{E}\left(X_f^{(K,\lambda)} X_g^{(K,\lambda)}\right) = \left\langle \tilde{K}^{1/2} f, \tilde{K}^{1/2} g \right\rangle_{L^2(\lambda)} = \langle f, Kg \rangle_{L^2(\lambda)},$$

where K is defined on a dense domain in $L^2(\lambda)$.

Moreover,

$$\rho^{(\lambda, K)}(A \times B) := \int_A K(x, B) \, \lambda(dx) = \int_B K(y, A) \, \lambda(dy) = \lambda(A \cap B).$$

Definition 4.50. Let $(X, \mathscr{B})$ be a measure space, $\mathscr{B}$ a specified σ-algebra of subsets of X. A transition kernel K on $X \times \mathscr{B}$ is such that $K(x, \cdot)$ is a positive measure on $(X, \mathscr{B})$ for every x. Given K, set

$$\mathfrak{M}(K) = \{\lambda \text{ positive measure on } (X, \mathscr{B}) \mid \lambda(dx) K(x, dy)$$

$$= \lambda(dy) K(y, dx)\}. \tag{4.51}$$

If $\lambda \in \mathfrak{M}(K)$ we say that K defines a reversible transition.

Question 4.51. (1) Is $\mathfrak{M}(K) \neq \{0\}$? (2) What is the nature of $\mathfrak{M}(K)$?

Summary.

(i) $K(A, B) = \lambda(A \cap B)$ on $\mathscr{B}_{fin} \times \mathscr{B}_{fin}$, where

$$\mathscr{B}_{fin} = \{A \in \mathscr{B} \mid \lambda(A) < \infty\}.$$

The corresponding RKHS $\mathscr{H}_\lambda$ consists of signed measures

$$F(A) = \tilde{f}(A) = \int_A f \, d\lambda$$

with

$$\|F\|_{\mathscr{H}_\lambda}^2 = \int |f|^2 \, d\mu < \infty.$$

(ii) $K(x, \cdot)$, as a measure in the second variable, induces a p.d. kernel on $\mathscr{B}_{fin} \times \mathscr{B}_{fin}$ by

$$K^{ind}(A, B) = \int_A \lambda(dx) K(x, B).$$

The corresponding RKHS is denoted $\mathscr{H}_{K,\lambda}$. Set

$$(Kf)(x) = \int K(x, dy) f(y);$$

then we have the transform

$$F(A) = \tilde{f}(A) = \int_A (Kf) d\lambda = \int K(x, A) f(x) d\lambda(x),$$

and

$$\|F\|^2_{\mathscr{H}_{K,\lambda}} = \int \left| \widetilde{K}^{1/2} f \right|^2 d\lambda,$$

where $\widetilde{K}$ is the Fredrichs extension of K, and $\widetilde{K}^{1/2}$ is defined by the spectral theorem.

4.2 Transforms and factorizations

Starting with a set S, and a positive definite function (kernel) K defined on $S \times S$, we then pass to the corresponding RKHS, $\mathscr{H}(K)$. Recall that $\mathscr{H}(K)$ is a Hilbert space of functions f on S, characterized by having the reproducing kernel property, defined from the inner product in $\mathscr{H}(K)$. Here we turn to a more detailed analysis of $\mathscr{H}(K)$: It is of interest to establish explicit conditions which will let us decide whether, or not, a given function f (defined on S) is in $\mathscr{H}(K)$. This is of significance since $\mathscr{H}(K)$, as a Hilbert space, is realized only as an abstract completion. The result below, Lemma 4.52, offers such a criterion. This lemma will also be used (in subsequent sections) below where our aim will be to offer explicit transforms (analogous to infinite-dimensional Fourier transforms) between suitable classes of L^2 spaces and $\mathscr{H}(K)$.

Lemma 4.52. *Let $K : S \times S \to \mathbb{R} \,(or\, \mathbb{C})$ be p.d., fixed; and let $\mathscr{H}(K)$ be the RKHS. A function F on S is in $\mathscr{H}(K)$ if and only if $\exists C = C_F < \infty$ such that $\forall n, \forall \{c_i\}_1^n, \{s_i\}_1^n, c_i \in \mathbb{R}, s_i \in S$, we have the estimate*

$$\left| \sum_i c_i F(s_i) \right|^2 \leq C \sum_i \sum_j c_i c_j K(s_i, s_j). \tag{4.52}$$

Summary of steps in the proof: Suppose F is a function on S such that (4.52) holds.

Step 1. Define

$$\psi_F \left(\sum_i c_k K(\cdot, s_i) \right) := \sum_i c_i F(s_i);$$

it is bounded $\mathscr{H}(K) \to \mathbb{R}$ by (4.52).

Step 2. Riesz applied to ψ_F, $\exists! H = H_F \in \mathscr{H}(K)$ such that

$$\psi_F(g) = \langle g, H \rangle_{\mathscr{H}(K)}.$$

Step 3. Set $g = K(\cdot, t)$, then $F(t) = H(t)$, so $F \in \mathscr{H}(K)$ since $H \in \mathscr{H}(K)$ by Riesz. Since ψ_F is $\mathscr{H}(K)$ bounded by (4.52) we may take the limit.

Proof. [Proof sketch] The non-trivial direction is $(4.52) \Longrightarrow F \in \mathscr{H}(K)$. So suppose F (fixed) is a function on S satisfying (4.52). Then define a linear functional $\psi = \psi_F :$ dense domain in $\mathscr{H}(K) \to \mathbb{R}$,

$$\psi\big(\underbrace{\sum_i c_i K(\cdot, s_i)}_{\in \mathscr{H}(K)}\big) = \sum_i c_i F(s_i) \in \mathbb{R} \tag{4.53}$$

(in the real case, also works over $\mathbb{C}$), and $\psi(K(\cdot, s)) = F(s)$, then $\psi = \psi_F$ is linear and bounded on $\mathscr{H}(K)$, and so extends by closure from $span\{K(\cdot, s)\}$ to all $\mathscr{H}(K)$. By Riesz, $\exists! \; H \in \mathscr{H}(K)$ such that

$$\psi(g) = \langle g, H \rangle_{\mathscr{H}(K)}, \quad \forall g \in \mathscr{H}(K).$$

Note $\psi = \psi_F$, and so $H = H_F$. Now take $g = K(\cdot, s)$,

$$\underbrace{\psi(K(\cdot, s))}_{=F(s) \text{ by (4.53)}} = \langle K(\cdot, s), H \rangle_{\mathscr{H}(K)} = H(s),$$

and so $F(s) = H(s)$ for all $s \in S$; since $H \in \mathscr{H}(K)$ by Riesz, we get $F \in \mathscr{H}(K)$. $\qquad \square$

Two applications. Fix $K : S \times S \to \mathbb{R}$ given p.d.

Application 1. Suppose $\{e_n(\cdot)\}_{n \in \mathbb{N}}$ is a system of functions on S such that

$$K(s, t) = \sum_{n \in \mathbb{N}} e_n(s) e_n(t), \tag{4.54}$$

then an application of Lemma 4.52 $\Longrightarrow e_n \in \mathscr{H}(K)$, $\forall n \in \mathbb{N}$.

Application 2. Suppose $(X, \mathscr{B}_X, \mu)$, μ positive measure, is a measure space such that

$$K(s, t) = \int_X \psi(x, s) \psi(x, t) \mu(dx) \tag{4.55}$$

holds, then for all $A \in \mathscr{B}_X$ with $\mu(A) < \infty$, we set

$$e_A(s) = \int_A \psi(s, x) \mu(dx). \tag{4.56}$$

Lemma 4.53. $\mu(A) < \infty \iff e_A \in \mathscr{H}(K)$.

Recall that $J = J_\psi : L^2(\mu) \to \mathscr{H}(K)$,

$$(Jf)(s) := \int_X f(x) \psi(x, s) \mu(ds),$$

and $J^* : \mathscr{H}(K) \to L^2(\mu)$. We prove that J^* is isometric, so $JJ^* = I_{\mathscr{H}(K)}$.

Also, if $A \in \mathscr{B}_X$, $\mu(A) < \infty$, then $\chi_A = f \in L^2(\mu)$, so the vectors $e_A(s) := \int_A \psi(x, s) \mu(dx)$ are special cases of the $L^2(\mu)$ analysis, i.e., $J(\chi_A) =: e_A \in \mathscr{H}(K)$. If ξ is a signed measure on S, $\xi K \xi < \infty$, then

$$J^*(T_K(\xi))(x) = \int_S \psi(x, s) \xi(ds), \quad x \in X.$$

In application 2, we fix $(X, \mathscr{B}_X, \mu, \psi(s, x))$ such that

$$K(s, t) = \int_X \psi(s, x) \psi(t, x) \mu(dx), \tag{4.57}$$

$\psi(s, \cdot) \in L^2(\mu)$, $\forall x \in S$. Define

$$e_A(s) = \int_A \psi(s, x) \mu(ds).$$

Note that (4.57) is a measurable version of (4.54). Note that in the estimates (4.52) to the two cases we get:

<u>CASE 1</u> $F = e_n$, n fixed, $\|e_n\|_{\mathscr{H}(K)} \leq 1$; and

<u>CASE 2</u> $F = e_A$ (see (4.56)), $\|e_A\|^2_{\mathscr{H}(K)} \leq \mu(A)$.

Lemma 4.54. *We have* $\|e_A\|^2_{\mathscr{H}(K)} \leq \mu(A)$, $\forall A \in \mathscr{B}$.

Proof. Let $e_A(s) = \int_A \psi(s, x) \mu(dx)$; pick n, $\{c_i\}_1^n$, $\{s_i\}_1^n$ and

$$\left| \sum_i c_i \int_A \psi(s_i, x) \mu(dx) \right|^2 = \left| \int_A \sum_i c_i \psi(s_i, x) \mu(dx) \right|$$

$$\overset{\text{(by Schwarz)}}{\leq} \mu(A) \int_X \left| \sum_i c_i \psi(s_i, x) \right|^2 \mu(dx)$$

$$= \mu(A) \sum_i \sum_j c_i c_j \underbrace{\int_X \psi(s_i, x) \psi(s_j, x) \mu(dx)}_{=K(s_i, s_j) \text{ by (4.57)}}$$

$$= \mu(A) \sum_i \sum_j c_i c_j K(s_i, s_j).$$

Hence, $\|e_A\|^2_{\mathscr{H}(K)} \leq \mu(A)$. $\qquad\qquad \square$

Properties and questions.

(i) Suppose $K(s,t) = \sum_{n \in \mathbb{N}} e_n(s) e_n(t)$, then for the e_n's (distinct)

$$\|e_n\|_{\mathscr{H}(K)} = 1, \ \forall n \iff \{e_n\} \text{ is an ONB of } \mathscr{H}(K).$$

In general we only have $\|e_n\|_{\mathscr{H}} \leq 1$.

(ii) Let $(X, \mathscr{B}_X, \mu)$ be such that $K(s,t) = \int_X \psi(s,x) \psi(t,x) \mu(dx)$. For $A \in \mathscr{B}_{fin}$, set

$$e_A(s) = \int_A \psi(s,x) \mu(dx).$$

We proved that $\|e_A\|^2_{\mathscr{H}(K)} \leq \mu(A)$.

Proof details.

Fix $K : S \times S \to \mathbb{R}$ p.d. and let $\mathscr{H}(K)$ be the RKHS; fix a measure space $(X, \mathscr{B}, \mu)$, μ positive σ-finite, and a function $\psi : X \times S \to \mathbb{R}$, such that

$$K(s,t) = \int_X \psi(x,s) \psi(x,t) \mu(dx).$$

Set

$$\mathscr{D}_{fin}(\mu) := span\{\chi_A : A \in \mathscr{B}_{fin}\},$$

so

$$(T_\psi f)(s) = \int f(x) \psi(x,s) \mu(dx), \quad f \in \mathscr{D}_{fin}(\pi). \qquad (4.58)$$

is coisometric, and that $J = J_\psi$ is defined as an operator $L^2(\mu) \to \mathscr{H}(K)$ with domain $= \mathscr{D}_{fin}$.

Lemma 4.55. *The operator* $J_\psi : L^2(\mu) \to \mathscr{H}(K)$ *is closable, and we have a symmetric pair as follows. Define* J_ψ^* *on* $\mathscr{H}(K)$ *by*

$$J_\psi^*\Big(\underbrace{\sum_i c_i K(\cdot, s)}_{\text{finite sum}}\Big)(x) := \sum_i c_i \psi(x, s_i); \tag{4.59}$$

note then for $\forall s,\ \psi(\cdot, s) \in L^2(X, \mu)$. *We have*

$$\langle J_\psi f, G \rangle_{\mathscr{H}(K)} = \langle f, J_\psi^*(G) \rangle_{L^2(\mu)}. \tag{4.60}$$

Note that J_ψ^* *is determined on* $\mathrm{span}\{K(\cdot, s) : s \in S\}$ *by*

$$J_\psi^*(K(\cdot, s))(x) = \psi(x, s) \tag{4.61}$$

Proof. Verification of (4.61): $f \in \mathscr{D}^{fin}(\mu)$:

$$
\begin{aligned}
\langle J_\psi f, K(\cdot, s) \rangle_{\mathscr{H}(K)} &= (J_\psi f)(s) \\
&\overset{\text{by (4.58)}}{=} \int_X f(x)\,\psi(x, s)\,\mu(dx) \\
&= \langle f(\cdot), \psi(\cdot, s) \rangle_{L^2(\mu)},
\end{aligned}
$$

and (4.61) follows. $\square$

Lemma 4.56.

(i) *The operator* $J_\psi : L^2(\mu) \to \mathscr{H}(K)$ *is contractive, i.e.,*

$$\|J_\psi f\|_{\mathscr{H}(K)} \leq \|f\|_{L^2(\mu)}, \quad \forall f \in L^2(\mu);$$

and also $J_\psi J_\psi^* = I_{\mathscr{H}(K)}$, *so* J_ψ *is coisometric.*

(ii) *We have for* $f \in L^2(\mu),\ s, t \in S$:

$$J_\psi J_\psi^*(K(\cdot, s))(t) = \int_X \psi(x, s)\,\psi(x, t)\,\mu(dx) = K(s, t).$$

(iii) $J_\psi^*(K(\cdot, s))(x) = \psi(x, s),\ \forall s \in S.$

Proof.

$$
\begin{aligned}
\Big|\sum_i c_i (J_\psi f)(s_i)\Big|^2 &= \Big|\sum_i c_i \int f(x)\,\psi(x, s_i)\,\mu(dx)\Big|^2 \\
&= \Big|\int f(x) \sum_i c_i \psi(x, s_i)\,\mu(dx)\Big|^2
\end{aligned}
$$

$$\overset{\text{(by Schwarz)}}{\leq} \left(\int |f|^2 \, d\mu \right) \int \left| \sum_i c_i \psi \, (x, s_i) \right|^2 \mu \, (dx)$$

$$= \|f\|^2_{L^2(\mu)} \sum_i \sum_j c_i c_j \int \psi \, (x, s_i) \, \psi \, (x, s_j) \, \mu \, (dx)$$

$$= \|f\|^2_{L^2(\mu)} \sum_i \sum_j c_i c_j K \, (s_i, s_j)$$

$$\implies \|J_\psi f\|_{\mathscr{H}(K)} \leq \|f\|_{L^2(\mu)}.$$

□

Corollary 4.57 (Integral Frames). *J_ψ maps* <u>onto</u> *$\mathscr{H}(K)$ so we get a frame representation*

$$H(s) = \int f(x) \, \psi \, (x, s) \, \mu \, (dx) \quad \text{for } \forall H \in \mathscr{H}(K).$$

Proof. Since J_ψ^* is isometric, $J_\psi \left(L^2(\mu) \right)$ is closed in $\mathscr{H}(K)$, so we only need to prove that $ker \left(J_\psi^* \right) = 0$; but if $J_\psi^* H = 0$, $\implies J_\psi J_\psi^* H = H = 0$, which is the desired conclusion. □

The coisometry property $J_\psi J_\psi^* = I_{\mathscr{H}(K)}$ may be stated as a Parseval frame identity, in a measure setting.

For $H \in \mathscr{H}(K)$, set $J_\psi^* H = f \in L^2(\mu)$, then

$$\|H\|^2_{\mathscr{H}(K)} = \left\langle H, J_\psi J_\psi^* H \right\rangle_{\mathscr{H}(K)} = \left\langle J_\psi^* H, J_\psi^* H \right\rangle_{L^2(\mu)}$$

$$= \int_X \left| \left(J_\psi^* H \right)(x) \right|^2 \mu \, (dx) = \int_X |f(x)|^2 \, \mu \, (dx),$$

which is the Parseval frame identity.

Corollary 4.58. *If ξ is a signed measure on $(S, \mathscr{B}_S)$ and $K : S \times S \to \mathbb{R}$ is fixed p.d., then we get J_ψ for every scalar ψ $(X, \mathscr{B}_X, \mu)$*

$$\int_X \psi(s, x) \psi(t, x) \mu(dx) = K(s, t),$$

and

$$(J_\psi \xi)(\cdot) = \int_S \xi(ds) K(s, \cdot) \in \mathscr{H}(K) \iff \xi K \xi < \infty;$$

then

$$\left(J_\psi^* J_\psi \xi \right)(x) = \int \xi(ds) \, \psi(s, x),$$

and an application of Fubini.

Corollary 4.59. *If $\{X_s\}_{s\in X}$ is a Gaussian process on $(\Omega, \mathscr{C}, \mathbb{P})$ such that $\mathbb{E}(X_s X_t) = K(s,t)$ then for $\forall f \in L^2(\Omega, \mathbb{P})$,*

$$J_\psi(f) = \mathbb{E}(f X_s), \quad \forall s \in S.$$

The transform maps onto $\mathscr{H}(K)$.

Proof. For $f \in L^2(\Omega, \mathbb{P})$, $H \in \mathscr{H}(K)$

$$\mathbb{E}(f J_\psi^* H) = 0, \forall f \implies J_\psi^* H = 0 \implies H = 0,$$

since $\|H\|_{\mathscr{H}(K)}^2 = \int |J_\psi^* H|^2 d\mathbb{P} = \mathbb{E}(|J_\psi^* H|^2) = 0.$ □

There are related transforms defined from the setting $K(s,t) = \mathbb{E}(X_s X_t)$ where $\{X_s\}$ is a Gaussian process, i.e., $X_s : \Omega \to \mathbb{R}$ is Gaussian; $(\Omega, \mathscr{C}, \mathbb{P})$ fixed.

For $f \in L^2(\Omega, \mathbb{P})$, consider three cases

$$\mathbb{E}(X_s f) =: (J_1 f)(s)$$
$$\mathbb{E}(f e^{X_s}) =: (J_2 f)(s)$$
$$\mathbb{E}(f e^{iX_s}) =: (J_3 f)(s)$$

In the general case of $K : S \times S \to \mathbb{R}$ p.d. there may be many representations $\{e_n\}$ frames on S such that

$$K(s,t) = \sum_{n \in \mathbb{N}} e_n(s) e_n(t), \tag{4.62}$$

but the functions $e_n(\cdot)$ may not be an ONB in $\mathscr{H}(K)$; but we always have that $e_n(\cdot) \in \mathscr{H}(K) \, \forall n$; and they form a Parseval frame.

Also note that (4.62) is a special case of the measure factorizations as in (4.55) above, that is, $(X, \mathscr{B}_X, \mu, \psi)$, with

$$K(s,t) = \int_X \psi(x,s) \psi(x,t) \mu(dx),$$

so that $(X, \mu) = (\mathbb{N}, \text{counting measure})$ is the special case.

Remark 4.60. In the special case $X = \mathbb{N}$, the computation

$$K(s,t) = \sum_{n \in \mathbb{N}} e_n(s) e_n(t) \implies e_n \in \mathscr{H}(K)$$

extends to integral factorizations as well.

Suppose $\exists\, (X, \mathscr{B}_X, \mu)$ such that $\psi(\cdot, s) \in L^2(\mu)$, $\forall s \in S$ and

$$K(s,t) = \int_X \psi(x,s)\,\psi(x,t)\,\mu(dx), \qquad (4.63)$$

then for $\forall A \in \mathscr{B}_X$ with $\mu(A) < \infty$ we have

$$s \longmapsto e_A(s) = \int_A \psi(x,s)\,\mu(dx) \in \mathscr{H}(K).$$

Proof. Same argument, but now

$$\left| \sum_i c_i \underbrace{\int_A \psi(x,s_i)\,\mu(dx)}_{e_A(s_i)} \right| \leq \mu(A) \sum_i \sum_j K(s_i, s_j).$$

$\square$

Example 4.61. Brownian motion, $K(s,t) = s \wedge t$, $s,t \in [0,1]$. Pick a sample space $(\Omega, \mathscr{C}, \mathbb{P})$, $\{X_t\}_{t \in [0,1]}$ which realizes Brownian motion so $\mathbb{E}(X_s X_t) = K = s \wedge t$. Since

$$\mathbb{E}(X_s X_t) = \int_\Omega X_s(w)\,X_t(w)\,d\mathbb{P}(w),$$

the factorization (4.63) holds.

So we calculate that, for $\forall A \in \mathscr{C}$, so $A \subset \Omega$, $s \longrightarrow \mathbb{E}(X_s \in A) \in \mathscr{H}(K)$. Set

$$e_A(s) := \mathbb{E}(X_s \in A), \quad e_A(\cdot) \in \mathscr{H}(K), \; \forall A \in \mathscr{C},$$

and we compute

$$\langle e_A, e_B \rangle_{\mathscr{H}(K)} = \mathbb{P}(A \cap B), \quad \forall A, B \in \mathscr{C}.$$

Recall that $\mathscr{C}$ is the standard cylinder σ-algebra of subsets of Ω, so events. In this case we have

$$\|e_A\|^2_{\mathscr{H}(K)} = \mathbb{P}(A), \quad \forall A \in \mathscr{C}.$$

Also $\mathbb{P}$ is a probability measure, so $\mathbb{P}(A) < 1\,(<\infty)$.

Summary. Assuming $K(s,t) = \sum_{n \in \mathbb{N}} e_n(s)\,e_n(t) \implies e_n \in \mathscr{H}(K)$, $\forall n$

(i) $\{e_n\}$ is a Parseval frame in $\mathscr{H}(K)$.
(ii) $\|e_n\|_{\mathscr{H}(K)} \leq 1$, $\forall n$.

(iii)
$$\mathscr{H}(K) = \left\{ F(s) = \sum_{n \in \mathbb{N}} \alpha_n e_n(s) : (\alpha_n) \in l^2 \right\}, \quad \|F\|^2_{\mathscr{H}(K)} = \sum_{n \in \mathbb{N}} |\alpha_n|^2,$$

but (α_n) is not unique (unless ONB).

(iv) $\{e_n\}$ is an ONB $\Longleftrightarrow$ $\|e_n\|_{\mathscr{H}(K)} = 1$, $\forall n$. Then, $VF = \left(\langle F, e_n \rangle_{\mathscr{H}(K)} \right) \in l^2$ is an isometry $\mathscr{H}(K) \to l^2(\mathbb{N})$, onto.

Example 4.62. In may examples of p.d. kernels K on $S \times S$, there is a natural class of functions $e_n(\cdot)$ on S such that $K(s,t) = \sum_{n \in \mathbb{N}} e_n(s) e_n(t)$.

(i) $K(s,t) = \frac{1}{1-st}$; $s, t \in (-1, 1)$; take $e_n(s) = s^n$.

(ii) $K(s,t) = \frac{1}{1-z\bar{w}}$; $z, w \in \mathbb{D} = \{z \in \mathbb{C}, |z| < 1\}$; take $e_n(z^n)$.

(iii) $K(s,t) = e^{st}$, $s, t \in \mathbb{R}$; take $e_n(s) = s^n/\sqrt{n!}$.

As an application of the lemma, we get the following representations of the corresponding RKHS, $\mathscr{H}(K)$:

(i) $\mathscr{H}(K) = \left\{ F(s) = \sum_{n=0}^{\infty} \alpha_n s^n : (\alpha_n) \in l^2(\mathbb{N}_0) \right\}$.

$$\sum_{n=0}^{\infty} \alpha_n s^n \longleftrightarrow \int_X f(x) \psi(x, s) \mu(dx)$$

$$\alpha \in l^2 \longleftrightarrow L^2(\mu) \ni f$$

(ii) $\mathscr{H}(K) = \left\{ F(z) = \sum_{n=0}^{\infty} \alpha_n z^n : (\alpha_n) \in l^2(\mathbb{N}_0) \right\}$

(iii) $\mathscr{H}(K) = \left\{ F(z) = \sum_{n=0}^{\infty} \alpha_n s^n/\sqrt{n!} : (\alpha_n) \in l^2(\mathbb{N}_0) \right\}$

For the $\mathscr{H}(K)$-norm, we will have

$$\|F\|^2_{\mathscr{H}(K)} = \sum_{0}^{\infty} |\alpha_n|^2 \longleftrightarrow \|F\|^2_{\mathscr{H}(K)} = \int |f|^2 d\mu \quad (J^*F = f),$$

in all the Examples.

General case. Starting with $K : S \times S \to \mathbb{R}$ p.d., and $F_\mu^G \in \mathscr{H}(K)$, it is difficult to decide for example, if $F \in \mathscr{H}(K)$, or the norm $\|F\|_{\mathscr{H}(K)}$ or $\langle F, G \rangle_{\mathscr{H}(K)}$. If we introduce $\mathfrak{M}_2(K)$, or $\mathscr{L}_2(K)$, then we set $T_K(\mu)(t) = \int \mu(ds) K(s,t) \in \mathscr{H}(K)$ if $\mu K \mu = \iint \mu(ds) K(s,t)\mu(dt) < \infty$, and $\|T_K\mu\|^2_{\mathscr{H}(K)} = \mu K \mu$. We illustrate this below.

In the three examples, the function $\{e_n\} \subset \mathscr{H}(K)$ in fact form an ONB; and $\langle e_n, e_m \rangle_{\mathscr{H}(K)} = \delta_{n,m}$. the easiest way to prove this is to compute $T_K\left(\delta_0^{(n)}\right)(t) = n! t^n$ where $T_K(\mu)(t) = \mu(s) K(s,t)$.

Example 4.63. $K = \frac{1}{1-st}$, $s, t \in (-1, 1)$, so if $n \neq m$,

$$\delta_0^{(n)} K \delta_0^{(m)} = \left\langle T_k\left(\delta_0^{(n)}\right), T_k\left(\delta_0^{(m)}\right)\right\rangle_{\mathscr{H}(K)} = n!m! \langle e_n, e_m\rangle_{\mathscr{H}(K)} = 0$$

if $n \neq m$.

General. Given $K : S \times S \to \mathbb{R}$ p.d. and assume functions $(e_n(\cdot))$ on S satisfying

$$K(s,t) = \sum_{n \in \mathbb{N}} e_n(s) e_n(t) \text{ (the only assumption)} \qquad (4.64)$$

then $e_n(\cdot) \in \mathscr{H}(K) = \text{RKHS}$.

Proof. Fix n, for ex $n = 1$. Must show for $\forall \{c_i\}_1^n$, $\forall \{s_i\}_1^n$, $\exists C < \infty$ such that

$$
\begin{aligned}
\left| \sum_i c_i e_1(s_i) \right|^2 &\leq \sum_i \sum_j c_i c_j e_1(s_i) e_1(s_j) \\
&\leq \sum_{n \in \mathbb{N}} \sum_i \sum_j c_i c_j e_n(s_i) e_n(s_j) \\
&\overset{\text{by } (4.64)}{=} \sum_i \sum_j c_i c_j \underbrace{\sum_{n \in \mathbb{N}} e_n(s_i) e_n(s_j)}_{K(s_i, s_j) \text{ by } (4.64)} \\
&= \sum_i \sum_j c_i c_j K(s_i, s_j),
\end{aligned}
$$

and so $e_1 \in \mathscr{H}(K)$ by the lemma below. ($f = e_1$) The desired conclusion now follows from an application of the following, Lemma 4.64. Indeed, since

$$\left| \sum c_i e_1(s_i) \right|^2 \leq \sum_i \sum_j c_i c_j K(s_i, s_j),$$

so $\|e_1\|_{\mathscr{H}(K)} \leq 1$.

From (4.64), $\infty > K(s,s) = \sum_n e_n(s)^2$, and so $(e_n(\cdot)) \in l^2(\mathbb{N})$, $\forall s$. $\square$

The argument below relies on our earlier Lemma 4.52. For reader-friendliness, we have included here a few details in order to make the context more clear.

Lemma 4.64 (See Lemma 4.52 for details). *A function f on S is in $\mathscr{H}(K) \iff \exists C < \infty$ such that $\forall (c_i), \forall (s_i),$*

$$\left| \sum_i c_i f(s_i) \right|^2 \leq C \sum_i \sum_j c_i c_j K(s_i, s_j),$$

and $\inf C = \|f\|_{\mathscr{H}(K)}^2.$

Let K be a positive definite function (kernel) on a fixed set, S, and let $\mathscr{H}(K)$ be the corresponding RKHS. The purpose of the following Lemma 4.65, and Corollaries 4.67 & 4.69, is to offer explicit representations (and transforms) for the elements in $\mathscr{H}(K)$. These representations will depend on choices made for factorizations of K.

Lemma 4.65.

$$\mathscr{H}(K) = \left\{ \sum_n \alpha_n e_n(\cdot) : (\alpha_n) \in l^2(\mathbb{N}) \right\}.$$

Proof. For

$$F(\cdot) = \sum_{n \in \mathbb{N}} \alpha_n e_n(\cdot), \tag{4.65}$$

set

$$\|F\|_{\mathscr{H}(K)}^2 := \sum_{n \in \mathbb{N}} |\alpha_n|^2; \tag{4.66}$$

so we have a new Hilbert system of functions on S, note

$$\left| \sum_n \alpha_n e_n(s) \right|^2 \underset{\text{Schwarz}}{\leq} \sum_n |\alpha_n|^2 \sum_n |e_n(s)|^2 < \infty, \tag{4.67}$$

so $\forall (\alpha_n) \in l^2$. The sum in (4.65) is well defined.

If $F(\cdot) = \sum_{n \in \mathbb{N}} \alpha_n e_n(\cdot)$, $G(\cdot) = \sum_{n \in \mathbb{N}} \beta_n e_n(\cdot)$, then (4.66) $\implies$

$$\langle F, G \rangle_{\mathscr{H}(K)} = \sum \alpha_n \beta_n; \tag{4.68}$$

$$\updownarrow$$

$$\langle F, G \rangle_{\mathscr{H}(K)} = \int f(x) g(x) \mu(dx)$$

and this is form an explicit of the Hilbert space $\mathscr{H}(K)$. We now show that it has the representation property; but if

$$F \in \mathscr{H}(K) \longleftrightarrow (\alpha_n) \in l^2$$

$$K(\cdot, t) \in \mathscr{H}(K) \longleftrightarrow (e_n(t)) \in l^2 \qquad (4.69)$$

by the previous argument. Now use the definition (4.68) for the inner product $\langle \cdot, \cdot \rangle_{\mathscr{H}(K)}$, so

$$\langle F, K(\cdot, t) \rangle_{\mathscr{H}(K)} \overset{\substack{\text{using } (4.68) \\ \text{see also } (4.69)}}{=} \sum_n \alpha_n e_n(t) \overset{\text{by } (4.65)}{=} F(t);$$

$F \longrightarrow (\alpha_n) \in l^2 : \langle \cdot, \cdot \rangle$, $K \longrightarrow e_n(t)$. So we have shown that $\mathscr{H}(K)$ has the property which defines the RKHS $\mathscr{H}(K)$, so it *is* "*the*" RKHS; moreover it is defined up to isometric isomorphism $\simeq$. □

Remark 4.66. We have to be careful in defining the Hilbert inner product in the space $\mathscr{H} : F(s) = \sum_{n \in \mathbb{N}} \alpha_n e_n(s)$, functions on S. Define

$$\|F\|_{\mathscr{H}}^2 = \sum_{n \in \mathbb{N}} |\alpha_n|^2. \qquad (4.70)$$

So with $G = \sum \beta_n e_n(\cdot)$, then

$$\langle F, G \rangle_{\mathscr{H}} = \frac{1}{4} \left(\|F + G\|_{\mathscr{H}}^2 - \|F - G\|_{\mathscr{H}}^2 \right) = \sum_{n \in \mathbb{N}} \alpha_n \beta_n. \qquad (4.71)$$

So we must proceed to a quotient in l^2.

So when $K = \sum_n e_n(s) e_n(t)$ is given, we get that $(e_n(s)) \in l^2(\mathbb{N})$, and so

$$\overline{span}\left(\{e_n(s)_n\}_{s \in S}\right)$$

is a subspace in $l^2(\mathbb{N})$, say $l^2(K) := \overline{span}\left(\{e_n(s)_n\}_{s \in S}\right)$, and of course

$$l^2(\mathbb{N}) \ominus l^2(K) = \left\{ (\alpha_n) \in l^2(\mathbb{N}) : \sum_n \alpha_n e_n(s) = 0, \, \forall s \in S \right\}.$$

In general, $\{e_n\}_{n \in \mathbb{N}}$ is only a *frame*, and so $l^2(K)$ will be a *proper* subspace in $l^2(\mathbb{N})$, and so $(\alpha_n) \to F(s) = \sum_n \alpha_n e_n(s) \in \mathscr{H}(K)$ has a kernel. If all we have is $K = \sum_n e_n(s) e_n(t)$ then there may be $(\alpha_n) \neq 0$ in l^2 such that

$\sum \alpha_n e_n(s) = 0 \ \forall s \in S.$ So

$$\mathcal{H}(K) \simeq l^2(\mathbb{N}) \, / \, \left\{ (\alpha_n) : \sum \alpha_n e_n(s) = 0 \ \forall s \right\}$$

$$= l^2(\mathbb{N}) \ominus \left((e_n(s))^{\perp} \right)_{s \in S}$$

$$= (e_n(s))^{\perp\perp},$$

and so $V : \mathcal{H} \to l^2$ may map onto $\overline{span}\left\{ \{e_n(s)_n\}_{s \in S} \right\}.$

Corollary 4.67. *Assume (4.64), i.e.,* $K = \sum_n e_n(s) e_n(t)$, *we conclude that* $\{e_n\}$ *is a Parseval frame, i.e.,*

$$\|F\|^2_{\mathcal{H}(K)} = \sum_n |\langle F, e_n \rangle|^2, \quad \forall F \in \mathcal{H}(K). \tag{4.72}$$

But in the lemma, we established an explicit form for $\mathcal{H}(K)$ *because we get* $\mathcal{H}(K)$ *in the form (4.65)–(4.67).*

Proof. The expression in (4.72) is well defined.

$$\text{LHS}_{(4.72)} = \sum_n |\alpha_n|^2$$

by (4.65)–(4.66). By (4.68), $\langle F, e_n \rangle_{\mathcal{H}(K)} = \alpha_n$, where

$$F \sim (\alpha_1, \alpha_2, \dots), \quad e_n \sim \big(0, 0, \dots, 0, \underset{n^{th}}{1}, 0, \dots \big),$$

and so

$$\text{RHS}_{(4.72)} = \sum_n |\alpha_n|^2 = \text{LHS}_{(4.72)}. \qquad \square$$

Corollary 4.68. *Assume (4.64), i.e.,* $K = \sum_n e_n(s) e_n(t)$, *then define* $V : \mathcal{H}(K) \to l^2(\mathbb{N})$ *by*

$$VF = \left(\langle F, e_n \rangle_{\mathcal{H}(K)} \right) \in l^2; \tag{4.73}$$

then V *is isometric, and*

$$V^*((\alpha_n))(s) = \sum_n \alpha_n e_n(s) \in \mathcal{H}(K). \tag{4.74}$$

Proof. Fix $F \in \mathcal{H}(K)$, then

$$\|VF\|^2_{\mathcal{H}(K)} = \sum \left| \langle F, e_n \rangle_{\mathcal{H}(K)} \right| = \|F\|^2_{\mathcal{H}(K)} \text{ by (4.72),} \quad \text{and}$$

$$\left\langle VF, \underbrace{(\beta_i)_{i\in\mathbb{N}}}_{\in l^2} \right\rangle_{l^2} = \sum_n \langle F, e_n(\cdot)\rangle_{\mathcal{H}(K)} \beta_n$$

$$= \left\langle F, \underbrace{\sum_n \beta_n e_n(\cdot)}_{V^*(\beta_i)} \right\rangle_{\mathcal{H}(K)} = \langle F, V^*(\beta_i)\rangle_{\mathcal{H}(K)}$$

and the desired conclusion (4.74) holds. □

General facts about V and V^*.

$V : \mathcal{H}(K) \to l^2$, $V^* : l^2 \to \mathcal{H}(K)$; see above. We proved that V is isometric, so $V^*V = I_{\mathcal{H}}$ or equivalently $V^*VF = F$, $V(\mathcal{H}(K)) = clspan^{l^2}((e_n(s))_{s\in S})$. But $V(\mathcal{H}(K)) \neq l^2(\mathbb{N})$ by the previous argument so $V^* = V^{-1}$ and V^* is also isometric so $VV^* \neq I_{l^2}$.

The proof of $F = \sum \langle F, e_n\rangle e_n$ can be done with the use of (4.72) so the Parseval property, so

$$\left\| F - \sum \langle F, e_n\rangle e_n \right\| = 0.$$

Non-uniqueness. So in the Parseval frame case, every $F \in \mathcal{H}(K)$ has this representation $F(s) = \sum_{n\in\mathbb{N}} \langle F, e_n\rangle_{\mathcal{H}(K)} e_n(s)$, but it may have other representations $\sum_{n\in\mathbb{N}} \alpha_n e_n(s)$, $(\alpha_i) \in l^2(\mathbb{N})$, $\alpha_n \neq \langle F, e_n\rangle_{\mathcal{H}(K)}$. In both cases we have

$$\|F\|^2_{\mathcal{H}(K)} = \sum_{n\in\mathbb{N}} \left| \langle F, e_n\rangle_{\mathcal{H}(K)} \right|^2 = \sum_{n\in\mathbb{N}} |\alpha_n|^2.$$

Corollary 4.69. *Given $K = \sum_n e_n(s)e_n(t)$, see (4.64), we get explicit representation*

$$F(s) = \sum \alpha_n e_n(s), \tag{4.75}$$

$(\alpha_n) \in l^2$, but the expansion in (4.75) is not unique unless (e_n) is an ONB in $\mathcal{H}(K)$.

Also note that (e_n) in (4.64) is an ONB $\iff \|e_n\|_{\mathcal{H}(K)} = 1 \ \forall n \in \mathbb{N}$.

Proof.

$$\|e_n\|^2_{\mathcal{H}(K)} = \langle e_n, e_n\rangle_{\mathcal{H}(K)} + \sum_{m\neq n} \left| \langle e_n, e_m\rangle_{\mathcal{H}(K)} \right|^2$$

so if $\|e_n\|_{\mathcal{H}(K)} = 1 \implies e_n \perp e_m \ \forall m \neq n$. □

Remark 4.70. The two V and V^* in Corollary 4.68 (see (4.73) & (4.74)) are well defined. In general, assume only (4.64), i.e., $K = \sum_n e_n(s) e_n(t)$ $\implies (e_n)$ is a Parseval frame, but V^* may not be isometric. Note if (δ_n) is an ONB in $l^2(\mathbb{N})$, $\delta_n(m) = \delta_{n,m}$, then $(V^*\delta_n)(s) = e_n(s)$, but if

$$\underbrace{\langle V^*\delta_n}_{e_n(\cdot)}, \underbrace{V^*\delta_n\rangle}_{e_m(\cdot)}{}_{\mathscr{H}(K)} = \langle \delta_n, \delta_m\rangle_{l^2} = \delta_{n,m},$$

it now follows that $\{e_n\}$ must be an ONB. But Parseval is generally not ONB.

Summary 4.71. Given (4.64), V in (4.73), V^* in (4.74), then $V^*V = I_{\mathscr{H}}$ (same as V being isometric), but $Q = VV^*$ is a non-trivial projection in $l^2(\mathbb{N})$, Q = projection onto $V(\mathscr{H}(K))$ as a closed subspace in $l^2(\mathbb{N})$. But

$$
\begin{array}{ccc}
 & V & \\
\mathscr{H}(K) & \overset{\displaystyle\frown}{\underset{\displaystyle\smile}{}} & Q(\mathscr{H}(K)) \qquad \subset l^2(\mathbb{N}) \\
 & \text{possible proper subspace of } l^2 & \\
 & V^* &
\end{array}
$$

Conditionally negative (CND) functions N are given for

$$Q(\varphi, \psi) = e^{-\frac{1}{2}N(\varphi-\psi)} = \mathbb{E}\left(e^{iX_\varphi - iX_\psi}\right), \tag{4.76}$$

X_φ Gaussian on $(\Omega, \mathscr{C}, \mathbb{P})$, $\mathbb{E}(\cdot) = \int_\Omega (\cdot) \, d\mathbb{P}$, $S \subset \mathscr{H}_N \subset S'$ Gelfand triple. Note that Q is positive definite,

$$N(\varphi - \psi) = \mathbb{E}\left(|X_\varphi - X_\psi|^2\right), \tag{4.77}$$

and we define the transform

$$J : L^2(\Omega, \mathbb{P}) \longmapsto \mathbb{E}\left(f e^{iX_\varphi}\right) \in \mathscr{H}(Q).$$

We study properties of the transform $T = T_N$,

$$T(\xi)(\varphi) := \mathbb{E}\left(f e^{iX_\varphi}\right), \ f \in L^2(\Omega, \mathbb{P}), \ \varphi \in S.$$

Theorem 4.72. *J is an isometric isomorphism of $J : L^2(\Omega, \mathbb{P}) \simeq \mathscr{H}(Q)$ onto.*

Proof.

$$J^*(Q(\varphi, \cdot))(w) = e^{iX_\varphi(w) - \frac{1}{2}N(\varphi)},$$

where $X_\varphi(w) = w(\varphi)$, $w \in S'$, $\Omega = S'$.

   We prove that T^* is isometric, but also that it maps onto $L^2(S', \mathbb{P})$. Reason: The exponentials $\{e^{iX_\varphi}\}_{\varphi \in S}$ span a dense subspace in $L^2(S', \mathbb{P})$.

For the isometry $J : L^2(S', \mathbb{P}) \to \mathcal{H}(Q)$ we have
$$J\left(e^{iX_\varphi}\right) = Q(\varphi, \cdot) e^{\frac{1}{2}N(\varphi)}, \quad \varphi \in S,$$
with $Q(\varphi, \psi) := e^{-\frac{1}{2}N(\varphi - \psi)}$. $\qquad\qquad\qquad\qquad\qquad\qquad\square$

Every Parseval frame induces a system of transition probabilities. Let $(v_k)_{i \in \mathbb{N}}$ be a Parseval frame in $\mathcal{H}$. For $u \in \mathcal{H}$,
$$\mathrm{Prob}(u \to v_k) = \langle u, v_k \rangle$$

$$\mathrm{Prob}(u \to w) = \sum_k \mathrm{Prob}(u \to v_k)\, \mathrm{Prob}(v_k \to w).$$

Or, in the measurable case,
$$K(s,t) = \int_X \psi(x,s)\, \psi(x,t)\, \mu(dx),$$
with $F \in \mathcal{H}(K)$, then
$$\mathrm{Prob}(\to A) = \int_A |(J^*F)(x)|^2 \, \mu(dx).$$

4.2.1 *Summary of Wiener processes*

Fix a sigma finite measure space $(X, \mathcal{B}_X, \nu)$, set $\mathcal{B}_X^{\#} = \{A \in \mathcal{B} \mid \nu(A) < \infty\}$. Then there are two *equivalent* realizations of the corresponding Wiener process $\xi = \xi^{(W)}$.

Case 1. Gaussian process indexed by $\mathcal{B}_X^{\#}$, $\xi_A^{(W)}$;

Case 2. Gaussian process indexed by $L^2(\nu)$,
$$\xi = \xi_f := \int f(x)\, d\xi_x^W \quad (\text{Itô integral}).$$

We assume mean zero $\mathbb{E}(\xi_A) = 0$, $\mathbb{E}(\xi_f) = 0$, so the process is determined by the covariance function $\mathbb{E}(\xi_A^{(W)} \xi_B^{(W)}) = \nu(A \cap B)$, or equivalently, $\mathbb{E}(\xi_f \xi_g) \overset{\text{Itô isometry}}{=} \langle f, g \rangle_{L^2(\nu)}$.

Then (R, ν) system and Gaussian process. We sketch below that (R, ν) satisfies the axioms above, then
$$\rho^{(R,\nu)}(A \times B) = \int K^{(R,\nu)}(A, B) := \int_A R(x, B)\, \nu(dx) = \lambda(A \times B),$$
$$\tag{4.78}$$
extend to be defined on $X \times X$, for all $A, B \in \mathcal{F}^{\#} := \{A : \nu(A) < \infty\}$, is positive definite. Given K p.d., by general theory, $\exists$ a Gaussian process $\{\xi_A^{(K)}\}_{A \in \mathcal{F}^{\#}}$ and a probability space $(\Omega, \mathcal{C}, \mathbb{P})$ such that
$$\mathbb{E}(\xi_A^{(K)} \xi_B^{(K)}) = K(A, B). \tag{4.79}$$
It will be interesting to study the properties of this Gaussian process in more detail.

White noise analysis.

Remark 4.73. In the case when R defines the in $L^2(\nu)$, i.e., $K(A, B) = \nu(A \cap B)$, then $\{\xi_A^W\}_{A \in \mathscr{F}^\#}$ is the corresponding white noise process $\nu \mapsto \xi^{W,\nu} =: \xi^W$, and

$$\mathbb{E}(\xi_A^W \xi_B^W) = \nu(A \cap B). \tag{4.80}$$

This is a basic Gaussian process; called the ν-Wiener process. We may also index $\xi^{(W)}$ by $L^2(\nu)$, so $\{\xi_f\}_{f \in L^2(\nu)}$ is a Gaussian process, but now

$$\mathbb{E}(\xi_f^W \xi_g^W) = \langle f, g \rangle_\nu = \int f(x) g(x) \nu(dx), \tag{4.81}$$

where $\xi_f^W = \int f(x) d\xi_x^W$ is a Wiener–Itô integral.

Discussion of (4.80)–(4.81).

Fix $(X, \mathscr{B}_X, \nu)$ the corresponding Gaussian process $\xi^{(W)}$ with $\mathbb{E}(\xi_A^W) = 0$, $\forall A$, and $\mathbb{E}(\xi_A^W \xi_B^W) = \nu(A \cap B)$, $\forall A, B \in \mathscr{B}_X$, is fundamental, called the Wiener(–Itô) process.

If $f \in L^2(\nu)$, then the integral $\int f(s) \xi_{dx}^W =: \xi_f^W$ is well defined, it is called the Itô-integral and we have $\mathbb{E}(|\xi_f^W|^2) = \int |f|^2 d\nu$, called the Itô-isometry. It is equivalent to isometry

$$\{\xi_A^W\}_{A \in \mathscr{B}_X} \longleftrightarrow \{\xi_f^W\}_{f \in L^2(\nu)}.$$

We have the operator Friedrichs extension $\widehat{R}$ (see, e.g., Remark 1.71 and [DS88]), and

$$\mathbb{E}(\xi_f \xi_g) = \langle \widehat{R}^{1/2} f, \widehat{R}^{1/2} g \rangle_\nu \tag{4.82}$$
$$= \mathbb{E}(\xi_{\widehat{R}^{1/2} f}^W \xi_{\widehat{R}^{1/2} g}^W).$$

So $\xi_f = \xi_{\widehat{R}^{1/2} f}^W$. But $R = PQ$, $\nu_1 = \nu_c = \nu$, and we get

$$\xi_f^{(P)} = \xi_{Qf}^W. \tag{4.83}$$

Recall that $\widehat{f}(A) = \int R(x, A) f(x) \nu(dx)$ is the transform

$$\mathscr{H}^{(R,\nu)}(K) \ni f \iff f \in dom(\widehat{R}^{1/2}).$$

Lemma 4.74. *Given* (R, ν) *and define*

$$\lambda(A \times B) = K_\lambda(A, B), \quad A, B \in \mathscr{B}_X$$
$$= \int_A \nu(dx) R(x, B) \quad (symmtric),$$

then $\exists$ *a Gaussian process* η_A *indexed by* $A \in \mathscr{B}^{\#}$ $(= \{A \in \mathscr{B} : \nu(A) < \infty\})$ *such that* K_λ *in* (4.83) *is the covariance function of* η_A, *i.e.,*

$$\mathbb{E}(\eta_A \eta_B) = K_\lambda(A, B), \quad \forall A, B \in \mathscr{B}^{\#}.$$

We may take

$$\xi_A := \xi^W_{Q(\cdot, A)}; \tag{4.84}$$

see proof below.

Proof. Let ξ^W be the Wiener process for $L^2(\nu)$, i.e.,

$$\mathbb{E}(\xi^W_f \xi^W_g) = \langle f, g \rangle_\nu = \int_X fg \, d\nu. \tag{4.85}$$

Consider a R, P, Q system for ν as usual, $\nu = \nu_1 = \nu_2$, then

$$Q(\cdot, A) \in L^2(\nu)$$
$$P(\cdot, A) \in L^2(\nu), \quad \forall A \in \mathscr{F}^{\#} = \{\nu(A) < \infty\}.$$

Set

$$\xi_A = \xi^W_{Q(\cdot, A)}; \tag{4.86}$$

it is Gaussian and then we get:

$$
\begin{aligned}
\mathbb{E}(\eta_A \eta_B) &= \mathbb{E}(\xi^W_{Q(\cdot, A)} \xi^W_{Q(\cdot, B)}) \\
&= \langle Q(\cdot, A), Q(\cdot, B) \rangle_{L^2(\nu)} \\
&\overset{\text{by (4.85)}}{=} \int Q(y, A) Q(y, B) \nu(dy) \\
&\overset{\text{by the lemma}}{=} \int_A R(x, B) \nu(dx) \\
&= K^{R,\nu}(A, B)
\end{aligned}
\tag{4.87}
$$

by the general lemma, $R = PQ$ from earlier. $\qquad \square$

Conclusion. When (R, ν) is given, then the Gaussian process corresponding to $K^{R,\nu}(\cdot, \cdot)$ is specified by (4.86). Note, we get a solution to $\mathbb{E}(\eta_A \eta_B) = K^{R,\nu}(A, B)$ as in (4.87) for every factorization $R = PQ$.

Th isometric isomorphism $dom(\widehat{R}^{1/2}) \longleftrightarrow \mathscr{H}_\lambda$ (RKHS). When (R, ν) are given, set

$$\lambda := \lambda^{(R,\nu)\prime\prime} = \int_A R(x, B)\nu(dx), \quad A, B \in \mathscr{B}_X.$$

$\lambda^{(R,\nu)}$ is a symmetric measure on $X \times X$. Given $f \in dom(\widehat{R}^{1/2})$, set

$$\widetilde{f}(A) := \int R(x, A) f(x)\nu(dx), \; A \in \mathscr{B}_X^\# \quad \text{(i.e., } \nu(A) < \infty). \quad (4.88)$$

To show that the *transform* really does the job: If $f \mapsto \widetilde{f}$ is defined as in (4.88), then

$$\widetilde{f} \in \mathscr{H}_\lambda \longleftrightarrow f \in dom(\widehat{R}^{1/2}). \quad (4.89)$$

(Here $\widehat{R}$ = the Friedrichs extension.)

Step 1. $\Leftarrow$ Suppose $f \in dom(\widehat{R}^{1/2})$, then $\forall a, \forall (c_i), \forall (A_i)$, $A_i \in \mathscr{B}^\#$, set $\varphi := \sum_i c_i \chi_{A_i}$ then

$$\sum_i c_i \widetilde{f}(A_i) = \sum c_i \langle R(\cdot, A_i), f \rangle_\nu = \langle R\varphi, f \rangle_\nu,$$

if $\varphi = \sum_i c_i \chi_{A_i}$.

4.2.2 Generalized Carleson measures

Let K be a positive definite function (kernel) on a fixed set, S, and let $\mathscr{H}(K)$ be the corresponding RKHS. As discussed the marginal functions from K define a cylinder sigma-algebra on S. A positive sigma-finite measure μ on S will be said to be generalized Carleson, if $\mathscr{H}(K)$ is boundedly contained in $L^2(S, \mu)$. The generalized Carleson measures will be denoted $GC(K)$.

While the introduction of positive kernels K do not make reference to a measure μ, it turns out that, given K, there is a rich variety of choices of measures μ, each one serving a particular purpose. The measures μ considered below, starting with Lemma 4.75, allows us to realize K as an integral kernel for a bounded operator, say T_K acting in $L^2(\mu)$. See also Proposition 4.85 below for additional details.

Lemma 4.75. *The following two conditions for a signed measure μ on $(S, \mathscr{B}_{(S,K)})$ are equivalent:*

(i) *the function* $s \longmapsto \int K(s,t)\,\mu(dt) \in \mathscr{H}(K)$;

(ii) $\int\int \mu(dt)\,K(s,t)\,\mu(dt) < \infty$; *abbreviated* $\mu K \mu < \infty$.

Proof. (i) $\Rightarrow$ (ii) is easy. If (i) holds, set

$$K_\mu(\cdot) = \int K(\cdot,t)\,\mu(dt), \tag{4.90}$$

so $K_\mu \in \mathscr{H}(K)$, and compute $\|\cdot\|_{\mathscr{H}(K)}^2$:

$$\|K_\mu\|_{\mathscr{H}(K)}^2 = \left\langle \int K(\cdot,s)\,\mu(ds), \int K(\cdot,t)\,\mu(dt) \right\rangle_{\mathscr{H}(K)} \tag{4.91}$$

$$= \int\int \langle K(\cdot,s), K(\cdot,t)\rangle_{\mathscr{H}(K)}\,\mu(ds)\,\mu(dt)$$

$$= \int\int K(s,t)\,\mu(ds)\,\mu(dt) < \infty,$$

so (ii) holds.

(ii) $\Rightarrow$ (i) Assume (ii), and an application with simple functions $E_i \in \mathscr{B}_{(S,K)}$, $s_i \in E_i$, i.e., the usual filter created from partitions, and refinement.

Assuming (ii), we get the approximation

$$\sum_i \sum_j \mu(E_i)\,\mu(E_j)\,K(s_i,s_j) \xrightarrow[\text{limit over this filter}]{\text{LIMIT}} \mu K \mu = \text{the integral in (ii)}.$$

$$\tag{4.92}$$

But also note:

$$\left\| \sum_i \mu(E_i)\,K(\cdot,s_i) \right\|_{\mathscr{H}(K)}^2 = \sum_i \sum_j \mu(E_i)\,\mu(E_j)\,K(s_i,s_j),$$

and so the conclusion holds, $\sum_i \mu(E_i)\,K(\cdot,s_i) \to K_\mu$, $\|\cdot\|_{\mathscr{H}(K)}$ norm convergence, see (4.90). $\qquad\square$

A natural question and a partial answer

Assume $\mu K \mu < \infty$, then we get an induced p.d. kernel

$$K^{ind}(s,t) = \int_S K(s,x)\,K(t,x)\,\mu(dx).$$

Proposition 4.76. *With μ as above, then the following are equivalent:*

(i) $K^{ind} \ll CK$ *(ordering of p.d. kernels)*

(ii) $\mu \in GC(K)$, *generalized Carleson*

Proof. Compute

$$\int \left| \sum_i c_i K(s_i, x) \right|^2 \mu(dx) = \sum_i \sum_j c_i c_j K^{ind}(s_i, s_j).$$

$\square$

More general functions:

Lemma 4.77. *Let* $\xi : S \to \mathscr{L}$ *(Hilbert space) be a function such that* $K(s,t) = \langle \xi(s), \xi(t) \rangle_{\mathscr{L}}$, *then* $L(K_s) := \xi(s)$ *extends to an isometry,* $L : \mathscr{H}(K) \to \mathscr{L}$, *and for the adjoint* L^*,

$$L^*(f)(s) = \langle \xi(s), f \rangle_{\mathscr{L}}, \quad \forall f \in \mathscr{L}, \forall s \in S.$$

In particular, we have

$$s \longmapsto \langle \xi(s), f \rangle_{\mathscr{L}} \in \mathscr{H}(K).$$

Proof.

$$\langle K_s, L^* f \rangle_{\mathscr{H}(K)} = (L^* f)(s)$$
$$= \langle L K_s, f \rangle_{\mathscr{L}}$$
$$= \langle \xi(s), f \rangle_{\mathscr{L}}$$

which is the desired formula.

We have $L^* L = I_{\mathscr{H}(K)}$ and $LL^* = $ projection in $\mathscr{L}$ onto $L(\mathscr{H}(K))$. $\square$

4.2.3 Factorization of p.d. kernels

We now turn to certain applications, the special case when $\mathscr{L} = L^2(\mu)$.

Corollary 4.78. *Given* $K : S \times S \to \mathbb{R}$ *p.d. and* $(X, \mathscr{B}_X, \mu, \psi)$, *set*

$$K(s,t) = \int_X \psi(x, s) \psi(x, t) \mu(dx), \tag{4.93}$$

then there is a canonical isometry L *acting by extending* $L(K(\cdot, s)) := \psi(\cdot, s) \in L^2(\mu)$, *extend by linearity and closure.*

For the adjoint L^*, *we have,* $\forall f \in L^2(\mu) :$

$$L^*(f)(s) = \int_X f(x) \psi(x, s) \mu(dx) \tag{4.94}$$

$L^*(f) \in \mathscr{H}(K)$.

$$\mathscr{H}(K) \underset{L^*}{\overset{L}{\rightleftarrows}} L^2(\mu)$$

Since L is isometry, L^* is co-isometric, and therefore contractive:

$$\|L^*(f)\|_{\mathcal{H}(K)} \leq \|f\|^2_{L^2(\mu)} = \int_X |f|^2 \, d\mu.$$

Proof. Proof of (4.94)

$$\mathcal{H}(K) \xrightarrow[\text{isomtry}]{L} L^2(\mu) \ni f$$

$$K_s \xleftarrow{\quad L^* \quad} .$$

$$\langle L^* f, K_s \rangle_{\mathcal{H}(K)} = (L^* f)(s) \quad \text{the RKHS property}$$

$$= \langle f, \underbrace{LK_s}_{\psi(\cdot, s)} \rangle_{L^2(\mu)}$$

$$= \int_X f(x) \, \psi(x, s) \, \mu(dx),$$

that is

$$(L^* f)(s) = \int_X f(x) \, \psi(x, s) \, \mu(dx)$$

which is the desired Equation (4.94). $\qquad\square$

Remark 4.79. Note that the *factorization formula* (4.93) for a given p.d. kernel K makes precise the statement that K has a direct integral representation in terms of a measurable family of rank-one p.d. kernels; referring to the product expression inside the integral in (4.93). Similarly Equation (4.64) above makes precise a related, but different, assertion to the effect that K has a countable sum-representation, now in terms of a (different) system of rank-one p.d. kernels. One way to realize a solution to (4.64) is to pick an ONB for $\mathcal{H}(K)$. So, combining the two conclusions, we see that every p.d. kernel always has a variety of such factorizations; so both direct integral representations; and countable sum, representations. The only restriction on admissible measure spaces (M, μ) for allowing for solutions to (4.93) is that the Hilbert dimension of $L^2(\mu)$ is not smaller than that of the RKHS $\mathcal{H}(K)$.

Remark 4.80. Note that, if $(X, \mathcal{B}_X, \mu, \psi)$ satisfies

$$K(s, t) = \int_X \overline{\psi(x, s)} \psi(x, t) \, \mu(dx) \quad \text{(factorization)}$$

then, for all $f \in L^2(\mu)$, we have

$$s \to (L^* f)(s) = \int_X f(x) \overline{\psi(x,x)} \mu(dx) \in \mathscr{H}(K),$$

but $(s \mapsto \psi(s,x))$ may *not* be in $\mathscr{H}(K)$ for points $x \in X$. An example is the Szego kernel.

Example 4.81. $K(z,w) = \frac{1}{1-z\overline{w}}$, $z, w \in \mathbb{D}$ the disk. $X = [0, 2\pi)$ or the circle, $\psi(z,x) = \frac{1}{1-ze^{-ix}}$, then we have the factorization

$$\frac{1}{2\pi} \int_0^{2\pi} \psi(z,x) \overline{\psi(w,x)} dx = K(z,w),$$

and $\mathscr{H}(K) = \mathscr{H}_Z$.

Then the isometry L defined by extension in Hardy space.

$$H_2 \ni K(z, \cdot) \xrightarrow{\ \ L\ \ } \psi(z, \ \cdot \ ) \in L^2(0, 2\pi),$$
$$\underset{\mathbb{D}}{\downarrow} \qquad\qquad\qquad \underset{[0,2\pi)}{\downarrow}$$

where $L(H_2) \subset L^2(0, 2\pi)$, and $H_2 = \{\sum_{n=0}^{\infty} \alpha_n e^{inx} : (\alpha_n) \in l^2\}$, and $LL^* =$ the projection onto $L(H_2)$.

One may also obtain solutions $(X, \mathscr{B}_X, \mu, \psi)$ to the factorization problem when $K : S \times S \to \mathbb{R}$ is a given p.d. kernel.

Lemma 4.82. *Let* $K : S \times S \to \mathbb{R}$ *be given p.d. and suppose* (μ_2, ψ) *on* $(X, \mathscr{B}_X)$ *solution to the factorization problem*

$$K(s,t) = \int_X \psi(s,x) \psi(t,x) \mu_2(dx), \qquad \forall s, t \in S. \qquad (4.95)$$

If $\mu_2 \ll \mu_1$ *with Radon–Nikodym derivative* $m := d\mu_2/d\mu_1$; *then we also get a factorization for* $(\mu_1, \psi(s,x)\sqrt{m(x)})$.

Proof. We compute

$$\int_X \psi(s,x)\sqrt{m(x)} \psi(t,x)\sqrt{m(x)} \mu_1(dx)$$

$$= \int_X \psi(s,x)\psi(t,x) \underbrace{m(x)\mu_1(dx)}_{\mu_2(dx)}$$

$$= \int_X \psi(s,x)\psi(t,x) \mu_2(dx) = K(s,t)$$

where we used (4.95) in the last step. $\qquad\qquad\qquad\qquad\qquad\qquad \square$

Fix $S \times S \xrightarrow{K} \mathbb{R}$ p.d. Let

$$GC(K) = \{\text{positive measure } \mu : \mathscr{H}(K) \hookrightarrow L^2(\mu) \text{ bounded, i.e., } \exists C < \infty$$

$$\text{s.t. } \int |F(x)|^2 \mu(dx) \le C \|F\|_{\mathscr{H}(K)}, \ \forall F \in \mathscr{H}(K)\}.$$

Theorem 4.83. *For $\forall \mu \in GC(K)$, we have*

$$\{\varphi d\mu : \varphi \in L^2(\mu)\} \subseteq \mathfrak{M}_2(K).$$

The converse holds too.

Definition 4.84. Given $S \times S \xrightarrow{K} \mathbb{R}$ p.d. holds. Let S be given the Borel σ-algebra for the induced metric on S. A positive measure μ on S is called *generalized Carleson* if $\exists C < \infty$ such that $\int |F(x)|^2 \mu(dx) \le C \|F\|_{\mathscr{H}(K)}$, for all $F \in \mathscr{H}(K)$. Note this states that $\mathscr{H}(K)$ is boundedly contained in $L^2(S, \mu)$. Notation: $\mu \in GC(K)$.

Proposition 4.85. *If K and $\mathscr{H}(K)$ are as above and if μ is a generalized Carleson measure, then we can calculate the adjoint b^* to the induced mapping $\mathscr{H}(K) \xrightarrow{b} L^2(\mu)$.*

$$
\mathscr{H}(K) \underset{\underset{b}{\longrightarrow}}{\overset{b^*}{\longleftarrow}} L^2(\mu)
$$
$$\ni F \qquad\qquad \varphi \in$$

We have for $\forall \varphi \in L^2(\mu)$

$$(b^*\varphi)(s) = \int K(s, x) \varphi(x) \mu(dx). \tag{4.96}$$

Proof. For $F \in \mathscr{H}(K)$, $\varphi \in L^2(\mu)$, we have

$$
\langle bF, \varphi \rangle_{L^2(\mu)} = \int F(x) \varphi(x) \mu(dx)
$$

$$
= \int \langle F(\cdot), K(\cdot, x) \rangle_{\mathscr{H}(K)} \varphi(x) \mu(dx)
$$

$$
= \left\langle F(\cdot), \int K(\cdot, x) \varphi(x) \mu(dx) \right\rangle_{\mathscr{H}(K)},
$$

and the conclusion (4.96) follows. ☐

Corollary 4.86. *Let K be p.d. on $S \times S$, and let μ be a generalized Carleson measure, then for $\forall \varphi \in L^2(\mu)$, the signed measure $\varphi d\mu \in \mathfrak{M}_2(K)$, i.e.,*

$$\iint_{S \times S} \varphi(x) \mu(dx) K(x, y) \varphi(y) \mu(dy) < \infty, \tag{4.97}$$

so $\mu \in GC(K) \implies \{\varphi d\mu\} \subseteq \mathfrak{M}_2(K)$.

Proof. We show that if $\varphi \in L^2(\mu)$, $\mu \in GC(K)$, then $\int K(\cdot, y) \varphi(y) \mu(dy) =: K_\varphi(\cdot) \in \mathscr{H}(K)$. From this we get $\|F_\varphi\|_{\mathscr{H}(K)}^2 = \text{LHS}_{(4.97)}$. ☐

Theorem 4.87. *The converse holds too: Given* $S \times S \xrightarrow{K} \mathbb{R}$ *p.d. Then*

$$\mu \in GC(K) \tag{4.98}$$

$$\Updownarrow$$

$$\{\varphi d\mu\} \subseteq \mathfrak{M}_2(K). \tag{4.99}$$

Proof. We only need to prove $\Uparrow$. But if (4.99) holds, then $T_K(\varphi d\mu)(s) = \int K(s,t)\varphi(t)\mu(dt)$ is bounded $L^2(\mu) \xrightarrow{T} \mathscr{H}(K)$, and so $T_K^*(\mathscr{H}(K)) \hookrightarrow L^2(\mu)$ is also bounded. But $(T_K^*(H))(s) = H(s)$, by assumption. $\qquad \square$

Corollary 4.88. *When* $\mu \in GC(K)$, *then we introduce the bounded operators* b *and* b^* *and* $bb^* : L^2(\mu) \to L^2(\mu)$ *is then selfadjoint and bounded integral operator*

$$(bb^*\varphi)(s) = \int_S K(s,t)\varphi(t)\mu(dt), \ \forall \varphi \in L^2(\mu).$$

In general, the operator may not be compact, but it will be under more stringent assumptions placed on μ, *as application to just* $\mu \in GC(K)$.

4.3 Numerical models

> "*The most fruitful areas for the growth of the sciences are those which have been neglected as a no-man's land, between the various established fields.*" Norbert Wiener.

While successful numerical models abound in mathematics, we shall focus here on four interrelated such frameworks, and each directly connected to the present main themes.

The four models are: Complex-valued Gaussian processes; Hermite polynomials; realizations for the simple harmonic oscillator; and the Segal–Bargmann[1] transforms.

The numerical aspects of complex-valued Gaussian processes enter via their use in stochastic approximation. Here we focus on schemes for generating Monte Carlo simulations, and associated Karhunen–Love (KL) expansions (for details, see, e.g., Lemma 4.40 and Theorem 4.14). In summary, KL expansions are representations of a stochastic processes as infinite linear combinations of i.i.d. standard Gaussians. The i.i.d. $N(0,1)$s may be realized with number generators, i.e., computer generated numbers sampled

---

[1]See the short bios in Appendix A.

from the $N(0,1)$ distribution. The latter in turn correspond to choices of orthogonal functions. So KL expansions are analogous to Fourier series expansions.

The Karhunen–Loève expansion is also known as a Hotelling transform, or an eigenvector transform. It is further closely related to principal component analyses (PCA), technique widely used in image processing and in data analysis.

The other two themes, Hermite polynomials and harmonic oscillator analysis, are directly connected with algorithms for generating orthogonal systems of functions. The Hermite polynomials and the Hermite functions play a key role in the study of CCR representations, and in stochastic calculus. They are also closely connected to analysis of the Segal–Bargmann transform. While we shall present this transform here in its finite-dimensional variant, it was developed first by I. E. Segal in the infinite-dimensional context of quantum fields.

4.3.1 *Complex-valued Gaussian processes*

Fix a probability space $L^2(\Omega, \mathscr{B}, \mathbb{P})$. Let $X \in L^2(\mathbb{P})$ be a Gaussian random variable with mean 0 and variance σ. The corresponding generating function is defined as

$$\varphi(s) = \int_\Omega e^{sX} d\mathbb{P} = \int_\mathbb{R} e^{sx} d\mu(x) = e^{\frac{1}{2}\sigma s^2},$$

where $\mu := \mathbb{P} \circ X^{-1}$, and $d\mu/dx = \frac{1}{\sqrt{2\pi\sigma}} e^{-\frac{x^2}{2\sigma}}$.

Expanding φ into power series in s,

$$\varphi(s) = \sum_{n=0}^{\infty} \left(\int_\Omega X^n d\mathbb{P} \right) \frac{s^n}{n!} = \sum_{k=0}^{\infty} \left(\frac{(2k)!}{2^k k!} \sigma^k \right) \frac{s^{2k}}{(2k)!},$$

and it follows that

$$\int_\Omega X^n d\mathbb{P} = \begin{cases} 0 & n \text{ is odd} \\ \frac{(2m)!}{m! 2^m} \sigma^m & n = 2m \end{cases}.$$

Definition 4.89. We say $X_1, X_2, \ldots, X_n$ is a joint Gaussian system if, for all $t \in \mathbb{R}^n$,

$$\int_\Omega e^{t \cdot X} d\mathbb{P} = e^{\frac{1}{2} \langle t, At \rangle_{\mathbb{R}^n}},$$

where $A = (\langle X_i, X_j \rangle)_{i,j=1}^n$ is the covariance matrix, i.e.,

$$a_{i,j} = \int_\Omega X_i X_j d\mathbb{P}.$$

Hence, the joint probability density is

$$f_{X_1 \cdots X_n}(x) = (2\pi)^{-\frac{n}{2}} \det A^{-1} e^{-\frac{1}{2} \langle x, A^{-1} x \rangle_{\mathbb{R}^n}}.$$

Lemma 4.90. *The following are equivalent.*

(i) *The system $X_1, \cdots, X_n$ is jointly Gaussian.*
(ii) *For all $c = (c_j) \in \mathbb{R}^n$, $c_1 X_1 + \cdots + c_n X_n$ is Gaussian.*

Theorem 4.91. *For every positive definite (p.d.) kernel $K : X \times X \to \mathbb{C}$, there exists a complex-valued Gaussian process $\{V_x\}_{x \in X}$ satisfying*

$$\mathbb{E}\left[\overline{V_x} V_y\right] = K(x, y), \quad \mathbb{E}[V_x] = 0, \tag{4.100}$$
$$\mathbb{E}[V_x V_y] = 0. \tag{4.101}$$

(If K is real-valued, (4.101) is dropped. Otherwise, $K(x, y) = 0$, for all x, y.)

Corollary 4.92. *The following are immediate from (4.100)–(4.101).*

(i) *If $x = y$, then*

$$\mathbb{E}\left[|V_x|^2\right] = K(x, x). \tag{4.102}$$

(ii) *For all $x \in X$, write $V_x = r_x + i t_x$, where $r_x = \Re\{V_x\}$, and $t_x = \Im\{V_x\}$. Then (4.101) is equivalent to*

$$\mathbb{E}[r_x r_y] = \mathbb{E}[t_x t_y], \quad \mathbb{E}[r_x t_y] = -\mathbb{E}[t_x r_y]. \tag{4.103}$$

(iii) *A special case of (4.103) is when $x = y$, then*

$$\mathbb{E}\left[r_x^2\right] = \mathbb{E}\left[t_x^2\right], \quad \mathbb{E}[r_x t_x] = 0. \tag{4.104}$$

Therefore, for every V_x, its real and imaginary parts have the same variance, and are stochastically independent.

(iv) *Finally,* $\mathbb{E}\left[V_x\right] = 0$ *is equivalent to*

$$\mathbb{E}\left[r_x\right] = \mathbb{E}\left[t_x\right] = 0. \qquad (4.105)$$

Hence, the two real-valued processes, $\{r_x\}$ *and* $\{t_x\}$*, have mean zero.*

Remark 4.93. By (4.103), K splits as follows:

$$K\left(x,y\right) = \mathbb{E}\left[\overline{V_x}V_y\right]$$

$$= \mathbb{E}\left[\left(r_x - it_x\right)\left(r_y + it_y\right)\right]$$

$$= \mathbb{E}\left[r_x r_y\right] + \mathbb{E}\left[t_x t_y\right] + i\left(\mathbb{E}\left[r_x t_y\right] - \mathbb{E}\left[t_x r_y\right]\right)$$

$$= 2\mathbb{E}\left[r_x r_y\right] + i2\mathbb{E}\left[r_x t_y\right]. \qquad (4.106)$$

This suggests that

$$\mathbb{E}\left[r_x r_y\right] = \frac{1}{2}\Re K\left(x,y\right), \qquad (4.107)$$

$$\mathbb{E}\left[r_x t_y\right] = \frac{1}{2}\Im K\left(x,y\right). \qquad (4.108)$$

Proof of Theorem 4.91. Suppose $K = A + iB$, where $A = \Re K$, and $B = \Im K$. Since $K\left(x,y\right) = \overline{K\left(y,x\right)}$, then A is symmetric and B is anti-symmetric.

For all $z = c + id \in \mathbb{C}^N$, we have

$$\sum_{k=1}^{N} \overline{z_k} z_l K\left(x_k, x_l\right) = \langle z, Kz\rangle_{\mathbb{C}^N}$$

$$= \langle c + id, \left(A + iB\right)\left(c + id\right)\rangle_{\mathbb{C}^N}$$

$$= \langle c, Ac\rangle_{\mathbb{C}^N} - \langle c, Bd\rangle_{\mathbb{C}^N} + \langle d, Ad\rangle_{\mathbb{C}^N} + \langle d, Bc\rangle_{\mathbb{C}^N}$$

$$= \left\langle \begin{bmatrix} c \\ d \end{bmatrix}, \begin{bmatrix} A & -B \\ B & A \end{bmatrix} \begin{bmatrix} c \\ d \end{bmatrix} \right\rangle_{\mathbb{R}^{2N}}. \qquad (4.109)$$

Here, we used the fact that $A^T = A$, and $B^T = -B$. Hence the matrix

$$\begin{bmatrix} A & -B \\ B & A \end{bmatrix} \qquad (4.110)$$

is p.d. on $\mathbb{R}^{2N}$.

By general theory, there exists a real-valued Gaussian process $\{(u_x, v_x)\}_{x \in X}$, with mean zero, and whose finite-dimensional distribution is determined by (4.110). Specifically, given $x_1, \ldots, x_N$ in X, then $\left\{\left(u_{x_j}, v_{x_j}\right)\right\}_{j=1}^{N}$ satisfies

$$\mathbb{E}\left[u_{x_j} u_{x_k}\right] = \mathbb{E}\left[v_{x_j} v_{x_k}\right] = A\left(x_j, x_k\right),$$

$$\mathbb{E}\left[u_{x_j} v_{x_k}\right] = -\mathbb{E}\left[v_{x_j} u_{x_k}\right] = -B\left(x_j, x_k\right).$$

Set $V_x = \frac{1}{\sqrt{2}} (u_x - iv_x)$, then

$$\mathbb{E}\left[\overline{V_x}V_y\right] = \frac{1}{2}\mathbb{E}\left[(u_x + iv_x)(u_y - iv_y)\right]$$

$$= \frac{1}{2}\mathbb{E}\left[u_x u_y + v_x v_y - iu_x v_y + iv_x u_y\right]$$

$$= A(x,y) + iB(x,y) = K(x,y),$$

which is the assertion.

Since B is anti-symmetric, its main diagonal entries are zeros. This is equivalent to $\mathbb{E}[u_x v_x] = 0$, for all $x \in X$. Hence, the real and imaginary parts of every V_x are independent.

Corollary 4.94. *Let $\mathscr{H}$ be a complex Hilbert space. Then there exists a complex-valued Gaussian process $\{V_x\}_{x \in \mathscr{H}}$, such that $\mathbb{E}[V_x] = 0$, and*

$$\mathbb{E}\left[\overline{V_x}V_y\right] = \langle x, y \rangle,$$

$$\mathbb{E}[V_x V_y] = 0, \quad \forall x, y \in \mathscr{H}.$$

Proof. Set $K(x_i, x_j) := \langle x_i, x_j \rangle_{\mathscr{H}}$, positive definite, and apply Theorem 4.91. $\qquad\square$

4.3.2 *Hermite polynomials*

Recall that $\exp\left(sx - \frac{1}{2}s^2\right)$ is the generating function of the Hermite polynomials in the x variable. Expanding this into power series in s, i.e.,

$$\exp\left(sx - \frac{1}{2}s^2\right) = \sum_{n=0}^{\infty} \frac{H_n(x)}{n!}s^n = \sum_{k=0}^{\infty}\sum_{l=0}^{\infty} \frac{x^k}{k!}\frac{(-1)^l}{2^l l!}s^{s+2l}, \quad n := 2 + 2l$$

then

$$H_n(x) = \sum_{m=0}^{\lfloor \frac{n}{2} \rfloor} \frac{(-1)^m n!}{2^m (n-2m)! m!} x^{n-2m}$$

$$x^n = \sum_{m=0}^{\lfloor \frac{n}{2} \rfloor} \frac{n!}{2^m (n-2m)! m!} H_{n-2m}(x).$$

Theorem 4.95. *Let H_n be the nth Hermite polynomial. Then,*

(i) $H'_n(x) = nH_{n-1}(x)$.

(ii) $H_{n+1}(x) = xH_n(x) - nH_{n-1}(x)$; *or equivalently,*

$$\left(x - \frac{d}{dx}\right) H_n(x) = H_{n+1}(x).$$

Note, $x - d/dx$ is the creation operator.

(iii) $H''_n(x) = xH'_n(x) - nH_n(x)$; *or,*

$$-H''_n(x) + xH'_n(x) = nH_n(x).$$

(iv) *The following identity holds*

$$H_n(x) = (-1)^n e^{\frac{1}{2}x^2} \left(\frac{d}{ds}\right)^n e^{-\frac{1}{2}x^2}.$$

Proof. (i) & (ii) Differentiate φ with respect to x and s, respectively.

(iii) It follows that

$$H'_{n+1}(x) = (n+1)H_n(x)$$

$$(xH_n(x))' = H_n(x) + xH'_n(x)$$

and so

$$(n+1)H_n(x) = H_n(x) + xH'_n(x) - H''_n(x).$$

This yields

$$-H''_n(x) + xH'_n(x) = nH_n(x).$$

(iv) Since $sx - \frac{1}{2}s^2 = (s-x)^2 + \frac{1}{2}x^2$, then

$$e^{sx - \frac{1}{2}s^2} = e^{\frac{1}{2}x^2} e^{-\frac{1}{2}(s-x)^2} = \sum_{n=0}^{\infty} \frac{H_n(x)}{n!} s^n.$$

Therefore,

$$H_n(x) = e^{\frac{1}{2}x^2} \left(\frac{d}{ds}\right)^n \Big|_{s=0} e^{-\frac{1}{2}(s-x)^2}$$

$$= (-1)^n e^{\frac{1}{2}x^2} \left(\frac{d}{dx}\right)^n e^{-\frac{1}{2}(s-x)^2} \Big|_{s=0}$$

$$= (-1)^n e^{\frac{1}{2}x^2} \left(\frac{d}{dx}\right)^n e^{-\frac{1}{2}x^2}.$$

$\square$

4.3.3 *Simple harmonic oscillator*

Consider the Schrödinger representation of the canonical commutation relation in $L^2(\mathbb{R})$. Let $p = -i\frac{d}{dx}$ and $q = x = $ multiplication by x, acting in $L^2(\mathbb{R})$. The operators p, q are defined on the common dense domain $\mathcal{S}$, the Schwarz space, and satisfy

$$[p, q] = -i.$$

$\left(-i\frac{d}{dx}(xf(x)) = -if(x) + x\left(-i\frac{d}{dx}\right)f.\right)$ Let h be the corresponding Hamiltonian (or the number operator),

$$h = \frac{1}{2}(p^2 + q^2 - 1)$$

$$= \frac{1}{2}\left(-(d/dx)^2 + x^2 - 1\right).$$

The operator h has the following decomposition:

$$h = \left(\frac{q - ip}{\sqrt{2}}\right)\left(\frac{q + ip}{\sqrt{2}}\right) = \left(\frac{x - d/dx}{\sqrt{2}}\right)\left(\frac{x + d/dx}{\sqrt{2}}\right).$$

Definition 4.96. Let a and a^* be the annihilation and creation operators, where

$$a = \frac{q + ip}{\sqrt{2}} = \frac{x + d/dx}{\sqrt{2}}, \quad a^* = \frac{q - ip}{\sqrt{2}} = \frac{x - d/dx}{\sqrt{2}},$$

and so

$$h = a^*a = \frac{1}{2}(p^2 + q^2 - 1).$$

Lemma 4.97. *The following hold:*

(i) $h = a^*a$, $aa^* = h + 1$, $[a, a^*] = 1$;
(ii) $[a, (a^*)^n] = n(a^*)^{n-1}$;
(iii) $[h, a^*] = a^*$;
(iv) $[h, a] = -a$.

Lemma 4.98. *Let Ω be the ground state, i.e., suppose $a\Omega = 0$, and $\|\Omega\| = 1$. Then:*

(i) *The ground state is given by*

$$\Omega_0 = \pi^{-\frac{1}{4}}e^{-\frac{1}{2}x^2}.$$

(ii) $a\,(a^*)^n\,\Omega = n\,(a^*)^{n-1}\,\Omega$.

(iii) $h\,(a^*)^n\,\Omega = n\,(a^*)^n\,\Omega$.

(iv) $\left\|(a^*)^n\,\Omega\right\|^2 = n!$.

(v) *For all* $n = 0, 1, 2, \ldots$, *let*

$$\Omega_n = \frac{1}{\sqrt{n!}}\,(a^*)^n\,\Omega.$$

The system $\{\Omega_n : n \in \mathbb{N}_0\}$ *is an orthonormal basis in* $L^2\,(\mathbb{R})$.

(vi) *Specifically, the eigenfunctions are given by*

$$\Omega_n = \frac{1}{\sqrt{n!}}H_n\left(\sqrt{2}x\right)\Omega_0$$

$$= \frac{1}{\sqrt{n!}}H_n\left(\sqrt{2}x\right)\pi^{-\frac{1}{4}}e^{-\frac{x^2}{2}}.$$

(vii) *With respect to the ONB* $\{\Omega_n\}$, *the operators* a, a^* *and* h *have the following infinite-matrix representations:*

$$a = \begin{bmatrix} 0 & 1 & & \\ & 0 & \sqrt{2} & \\ & & 0 & \sqrt{3} \\ & & & \ddots & \ddots \end{bmatrix}, \quad a^* = \begin{bmatrix} 0 & & & \\ 1 & 0 & & \\ & \sqrt{2} & 0 & \\ & & \sqrt{3} & 0 \\ & & & \ddots & \ddots \end{bmatrix},$$

$$h = \begin{bmatrix} 1 & & & \\ & 2 & & \\ & & 3 & \\ & & & 4 \\ & & & & \ddots \end{bmatrix}.$$

Summary 4.99. We have the following correspondences.

Fock	$L^2\,(dx)$	$L^2\,(\mu),\ \mu = $ Gaussian
$e^h = \sum_{n=0}^{\infty}\frac{h^n}{n!}$	$e^{\sqrt{2}x - \frac{1}{2}\|h\|^2}\Omega_0$	$e^{x - \frac{1}{2}\|h\|^2}$
$\left\|e^h\right\|_{\Gamma}^2 = e^{\|h\|^2}$	$\|\cdot\|_{L^2(dx)}^2 = e^{\|h\|^2}$	$\|\cdot\|_{L^2(\mu)} = e^{\|h\|^2}$
h^n	$H_n\left(\sqrt{2}x, \|h\|\right)\Omega_0$	$H_n\left(x, \|h\|\right)$

Let $f \in L^2\,(\mu)$, then

$$\int |f\,(x)|^2\,\frac{1}{\sqrt{2\pi}}e^{-x^2/2}dx = \int \left|f\left(\sqrt{2}x\right)\right|^2\frac{1}{\sqrt{\pi}}e^{-x^2}dx$$

$$= \int \left|f\left(\sqrt{2}x\right)\frac{1}{\sqrt[4]{\pi}}e^{-x^2/2}\right|^2 dx.$$

Lemma 4.100. *The operator* $W : f(x) \longmapsto f(\sqrt{2}x)\,\Omega_0(x)$ *is an isometric isomorphism from* $L^2(d\mu)$ *onto* $L^2(dx)$, *where* $\Omega_0(x) = \frac{1}{\sqrt[4]{\pi}}e^{-x^2/2}$ *is the ground state. The adjoint is then given by*

$$W^*g(x) = g\left(x/\sqrt{2}\right)/\Omega_0(x), \quad g \in L^2(dx).$$

Proof. A simple change of variable.   □

Lemma 4.101. *Let* a, a^*, h, W *be as above, then*

$$W^*aWf = f'(x)$$

and

$$W^*a^*Wf = xf(x) - f'(x).$$

Therefore, the number operator h *is realized in* $L^2(\mu)$ *as*

$$W^*a^*aWf(x) = xf'(x) - f''(x).$$

Proof. One checks that

$$\left(x + \frac{d}{dx}\right)f\left(\sqrt{2}x\right)\Omega_0(x) = xf\left(\sqrt{2}x\right)\Omega_0(x) + \sqrt{2}f'\left(\sqrt{2}x\right)\Omega_0$$

$$+f\left(\sqrt{2}x\right)\Omega_0' = xf\left(\sqrt{2}x\right)\Omega_0(x)$$

$$+\sqrt{2}f'\left(\sqrt{2}x\right)\Omega_0 - xf\left(\sqrt{2}x\right)\Omega_0$$

$$= \sqrt{2}f'\left(\sqrt{2}x\right)\Omega_0.$$

Therefore, $W^*aWf = f'(x)$. On the other hand,

$$\left(x - \frac{d}{dx}\right)f\left(\sqrt{2}x\right)\Omega_0(x) = xf\left(\sqrt{2}x\right)\Omega_0(x) - \sqrt{2}f'\left(\sqrt{2}x\right)\Omega_0$$

$$+xf\left(\sqrt{2}x\right)\Omega_0 = 2xf\left(\sqrt{2}x\right)\Omega_0(x)$$

$$-\sqrt{2}f'\left(\sqrt{2}x\right)\Omega_0.$$

□

Set $T = \frac{d}{dx}$ in $L^2(\mu)$ then

$$\int f'(x)\,g(x)\,e^{-\frac{1}{2}x^2}\,dx = \int f(x)\,(x - g'(x))\,e^{-\frac{1}{2}x^2}\,dx.$$

Hence

$$T^* = x - \frac{d}{dx}.$$

It follows that

$$T^*T = x\frac{d}{dx} - \left(\frac{d}{dx}\right)^2.$$

Corollary 4.102. *We have*

$$T = W^*aW, \ T^* = W^*a^*W, \quad and \quad N = T^*T.$$

where N is the number operator in $L^2(\mu)$.

4.3.4 Segal–Bargmann transforms

This RKHS sometimes called the Segal–Bargmann kernel/RKHS.[2] Reason: Segal first did the case of complex Hilbert space $\mathscr{H}$, so in a sense the analysis of Bargmann is the special case when the complex dimension of $\mathscr{H}$ is 1. In both cases, we get creation and annihilation operators, in the $\dim \mathscr{H} = 1$ case: creation = multiplication by z, and annihilation = d/dz.

In details, the Gaussian kernel

$$K(x, y) = e^{-\sigma \|x-y\|^2_{\mathbb{R}^d}}$$

has three additional properties of relevance to the discussion below:

(i) it is *stationary*, i.e., $K(x+z, y+z) = K(x, y)$, $\forall x, y, z \in \mathbb{R}^d$;

(ii) it allows *analytic continuation* $\mathbb{R}^d \to \mathbb{C}^d$, i.e., $K^{\mathbb{C}}(z, w) = e^{-\sigma \|z-w\|^2_{\mathbb{C}^d}}$, $z, w \in \mathbb{C}^d$, and

(iii) it is the Fourier transform of the generator of *the heat semigroup*, i.e., the Gaussian

$$g_t(x) = \frac{1}{\sqrt{2\pi t}} e^{\frac{-x^2}{2t}}, \quad x \in \mathbb{R}^d, t > 0, \tag{4.111}$$

$x^2 := x_1^2 + \cdots + x_d^2$; with

$$\hat{g}_t(\xi) = e^{\frac{-t\xi^2}{2}}, \quad \xi \in \mathbb{R}^d, \tag{4.112}$$

and $\xi^2 := \xi_1^2 + \cdots + \xi_d^2$.

---

[2]Segal was interested in representations of CCR($\mathscr{H}$), and physics. Then $\dim \mathscr{H}$ is infinite. Apparently Segal did his version (with abbreviated proofs) before Bargmann's paper, and Segal told me that he wasn't interested in $\dim \mathscr{H} = 1$. He said that he had told Bargmann about it but Bargmann apparently forgot that conversation.

With the use of (i)–(iii), one checks that the associated *Segal–Bargmann transform* is as follows:

$$L^2\left(\mathbb{R}^d\right) \ni f \longmapsto \left(\widetilde{e^{t\Delta}f}\right)(z) \in \text{RKHS}\left(e^{t\,z\cdot\overline{w}}\right) \tag{4.113}$$

where $\Delta := \sum_{i=1}^{d}\left(\frac{\partial}{\partial x_i}\right)^2$, and

$$e^{t\Delta}f = g_t * f, \quad t > 0,\ f \in L^2\left(\mathbb{R}^d\right), \tag{4.114}$$

and $*$ denoting convolution.

Note the kernel on the RHS in (4.113) is p.d. on $\mathbb{C}^d$. By "$\sim$" in (4.113) we refer to analytic continuation via (ii), i.e., for $f \in L^2\left(\mathbb{R}^d\right)$,

$$\left(\widetilde{e^{t\Delta}f}\right)(z) = (2\pi t)^{-\frac{d}{2}} \int_{\mathbb{R}^d} e^{-\frac{|z-x|^2}{2t}} f(x)\, dx, \tag{4.115}$$

and $dx = dx_1 \cdots dx_d$. We used that $g_t(\cdot)$ in (4.111) has an entire analytic extension from $\mathbb{R}^d$ to $\mathbb{C}^d$; the latter is used in (4.115).

The above kernel ($t > 0$ fixed)

$$K_t(z,w) := e^{t\,z\overline{w}}, \quad z, w \in \mathbb{C} \tag{4.116}$$

arises in many applications. A natural choice of factorization for (4.116) is

$$k_n^{(t)}(z) := \frac{t^{\frac{n}{2}}}{\sqrt{n!}} z^n, \quad z \in \mathbb{C}.$$

The kernel in (4.116) is called the *Bargmann kernel* with parameter t; or the *Segal–Bargmann kernel*, as it is a key tool in many of the applications.

The RKHS of (4.116) is the Hilbert space of all analytic functions F on $\mathbb{C}^d$ which are L^2 with respect to the Gaussian on $\mathbb{C}^d$, i.e.,

$$\int_{\mathbb{C}^d} e^{-\frac{t|z|^2}{2}} |F(z)|^2\, d\lambda_{\mathbb{C}^d}(z) < \infty. \tag{4.117}$$

In (4.117), $|z|^2 := |z_1|^2 + \cdots + |z_d|^2$, $z \in \mathbb{C}^d$, and $d\lambda_{\mathbb{C}^d}(z)$ denotes the Lebesgue measure on $\mathbb{C}^d \simeq \mathbb{R}^{2d}$.

For more details, we refer to [AFMP94, DHK13, HM08, KW17, Kra13]. The reader might find the following three books useful as general references for details, theory, fundamental examples, and applications: [Alp01], [Kra13] and [PR16].

4.4 Gaussian Hilbert space

> *"Anyone who attempts to generate random numbers by deterministic means is, of course, living in a state of sin."* [Reference to computer generated random numbers.] John von Neumann

The literature on Gaussian Hilbert space, white noise analysis, and its relevance to Malliavin calculus is vast; and we limit ourselves here to citing [BØSW04, AJL11, AJ12, VFHN13, AJS14, AJ15, AØ15, Jan97, Kat19, PR14, Ki1, LPW10, App09], and the papers cited there.

Setting and Notation.

$\mathscr{L}$: a fixed *real* Hilbert space
$(\Omega, \mathcal{F}, \mathbb{P})$: a fixed probability space
$L^2(\Omega, \mathbb{P})$: the Hilbert space $L^2(\Omega, \mathcal{F}, \mathbb{P})$, also denoted by $L^2(\mathbb{P})$
$\mathbb{E}$: the mean or expectation functional, where $\mathbb{E}(\cdot\cdot) = \int_\Omega (\cdot\cdot)\, d\mathbb{P}$

Definition 4.103. Fix a *real* Hilbert space $\mathscr{L}$ and a given probability space $(\Omega, \mathcal{F}, \mathbb{P})$. We say the pair $(\mathscr{L}, (\Omega, \mathcal{F}, \mathbb{P}))$ is a *Gaussian Hilbert space*. A *Gaussian field* is a linear mapping $\Phi : \mathscr{L} \longrightarrow L^2(\Omega, \mathbb{P})$, such that

$$\{\Phi(h) \mid h \in \mathscr{L}\}$$

is a Gaussian process indexed by $\mathscr{L}$ satisfying:

(i) $\mathbb{E}(\Phi(h)) = 0, \forall h \in \mathscr{L}$;
(ii) $\forall n \in \mathbb{N}, \forall l_1, \ldots, l_n \subset \mathscr{L}$, the random variable $(\Phi(l_1), \ldots, \Phi(l_n))$ is jointly Gaussian, with

$$\mathbb{E}(\Phi(l_i)\Phi(l_j)) = \langle l_i, l_j \rangle, \qquad (4.118)$$

i.e., $(\langle l_i, l_j \rangle)_{i=1}^n$ = the covariance matrix. (For the existence of Gaussian fields, see the discussion below.)

Remark 4.104. For all finite systems $\{l_i\} \subset \mathscr{L}$, set $G_n = (\langle l_i, l_j \rangle)_{i,j=1}^n$, called the Gramian. Assume G_n non-singular for convenience, so that $\det G_n \neq 0$. Then there is an associated Gaussian density $g^{(G_n)}$ on $\mathbb{R}^n$,

$$g^{(G_n)}(x) = (2\pi)^{-n/2} (\det G_n)^{-1/2} \exp\left(-\frac{1}{2}\langle x, G_n^{-1}x \rangle_{\mathbb{R}^n}\right) \qquad (4.119)$$

The condition in (4.118) assumes that for all continuous functions $f : \mathbb{R}^n \longrightarrow \mathbb{R}$ (e.g., polynomials), we have

$$\mathbb{E}(\underbrace{f(\Phi(l_1), \ldots, \Phi(l_n))}_{\text{real valued}}) = \int_{\mathbb{R}^n} f(x)\, g^{(G_n)}(x)\, dx; \qquad (4.120)$$

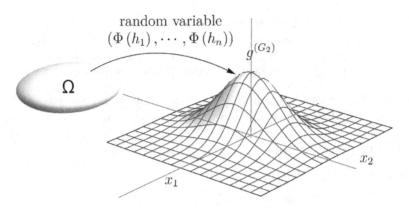

random variable
$(\Phi(h_1), \cdots, \Phi(h_n))$

$g^{(G_2)}$

Ω

x_2

x_1

Figure 4.8: The multivariate Gaussian $(\Phi(h_1), \cdots \Phi(h_n))$ and its distribution. The Gaussian with Gramian matrix (Gram matrix) G_n, $n = 2$.

where $x = (x_1, \ldots, x_n) \in \mathbb{R}^n$, and $dx = dx_1 \cdots dx_n =$ Lebesgue measure on $\mathbb{R}^n$. See Figure 4.8 for an illustration.

In particular, for $n = 2$, $\langle l_1, l_2 \rangle = \langle k, l \rangle$, and $f(x_1, x_2) = x_1 x_2$, we then get $\mathbb{E}(\Phi(k)\Phi(l)) = \langle k, l \rangle$, i.e., the inner product in $\mathcal{L}$.

For our applications, we need the following facts about Gaussian fields.

Fix a Hilbert space $\mathcal{L}$ over $\mathbb{R}$ with inner product $\langle \cdot, \cdot \rangle_{\mathcal{L}}$. Then (see [Hid80, AØ15, Gro70]) there is a *probability space* $(\Omega, \mathcal{F}, \mathbb{P})$, depending on $\mathcal{L}$, and a *real* linear mapping $\Phi : \mathcal{L} \longrightarrow L^2(\Omega, \mathcal{F}, \mathbb{P})$, i.e., a Gaussian field as specified in Definition 4.103, satisfying

$$\mathbb{E}(e^{i\Phi(k)}) = e^{-\frac{1}{2}\|k\|^2}, \quad \forall k \in \mathcal{L}. \tag{4.121}$$

It follows from the literature (see also [JT17]) that $\Phi(k)$ may be thought of as a generalized Itô-integral. One approach to this is to select a nuclear Fréchet space $\mathcal{S}$ with dual $\mathcal{S}'$ such that

$$\mathcal{S} \hookrightarrow \mathcal{L} \hookrightarrow \mathcal{S}' \tag{4.122}$$

forms a Gelfand triple. In this case we may take $\Omega = \mathcal{S}'$, and $\Phi(k)$, $k \in \mathcal{L}$, to be the extension of the mapping

$$\mathcal{S}' \ni \omega \longrightarrow \omega(\varphi) = \langle \varphi, \omega \rangle \tag{4.123}$$

defined initially only for $\varphi \in \mathcal{S}$, but, with the use of (4.123), now extended, via (4.121), from $\mathcal{S}$ to $\mathcal{L}$. See also Example 4.107 below.

Remark 4.105 (Joint distributions). In considerations of random processes, one must specify joint distributions. In particular, the requirement for a Gaussian process, is that all joint distributions must be Gaussian.

However for our present families of Gaussian processes, the requirement of joint Gaussian distributions is automatically built into the way we build Gaussian Hilbert spaces $\mathscr{H}$, or more precisely systems of centered Gaussian random variables, $\Phi(h)$, $h \in \mathscr{H}$. The reasoning is as follows: When starting with a Gaussian Hilbert spaces $\mathscr{H}$, or $\Phi(h)$, $h \in \mathscr{H}$; we must first refer to a probability space $(\Omega, \mathscr{C}, \mathbb{P})$, and so make precise the particular probability space which guarantees that $\mathscr{H}$, or $\Phi(h)$, $h \in \mathscr{H}$, is realized as a centered Gaussian process.

Indeed, to even define distributions, joint, or otherwise, we must first specify $(\Omega, \mathscr{C}, \mathbb{P})$, $\mathscr{C}$ the sigma-algebra of events, and $\mathbb{P}$ the probability measure. This can be done in any number of ways, for example, with a suitable choice of Gelfand triple; or alternatively with a direct use of the Kolmogorov inductive limit construction.

In more detail: How to get $(\Omega, \mathscr{C}, \mathbb{P})$? Pick your favorite Ω, set of sample-points, for example, Ω could be an infinite Cartesian product, a space of functions, or a choice of Schwartz tempered distributions (Section 4.1.2). Having made a choice, we then let $\mathscr{C}$ be the corresponding sigma-algebra generated by cylinder sets in Ω.

In the case of Gaussian Hilbert spaces $\mathscr{H}$, one may realize the probability measure $\mathbb{P}$ used in specification of the probability of events, as follows. We will build $\mathbb{P}$ as Kolmogorov inductive limit measure, inductive limit over the filter of all cylinders; each cylinder event is defined from a choice of a finite subset $\{h_i\}$ of vectors in $\mathscr{H}$. The corresponding limit leading to $\mathbb{P}$ will then be over all Gaussian marginals, specified s.t., the covariance matrix at each step, a finite subset $\{h_i\}$, will be the Gramian G, say size $n \times n$. The entries in G are the $\mathscr{H}$-inner products formed from pairs from the particular finite subset $\{h_i\}$ of $\mathscr{H}$. Specifically, $G = \{\langle h_i, h_j \rangle\}$ as an n-by-n matrix.

Conclusion. From this inductive-limit construction, leading to the probability measure $\mathbb{P}$, it will then follow that the joint distribution of every finite subsystem $\{\Phi(h_i)\}$ of random variables formed this way; will automatically be Gaussian, with covariance matrix equal to the n-dimensional Gramian matrix G, and so all joint distributions from the process $\{\Phi(h) : h \in \mathscr{H}\}$ are centered, Gaussian, and therefore specified by covariance matrices arising this way as Gramians G.

4.4.1 *White noise analysis*

Example 4.106. Fix a measure space $(X, \mathcal{B}, \mu)$. Let $\Phi : L^2(\mu) \longrightarrow L^2(\Omega, \mathbb{P})$ be a Gaussian field such that

$$\mathbb{E}(\Phi_A \Phi_B) = \mu(A \cap B), \quad \forall A, B \in \mathcal{B}$$

where $\Phi_E := \Phi(\chi_E)$, $\forall E \in \mathcal{B}$; and χ_E denotes the characteristic function. In this case, $\mathscr{L} = L^2(X, \mu)$.

Then we have $\Phi(k) = \int_X k(x) \, d\Phi$, i.e., the Itô-integral, and the following holds:

$$\mathbb{E}(\Phi(k) \Phi(l)) = \langle k, l \rangle = \int_X k(x) l(x) \, d\mu(x) \qquad (4.124)$$

for all $k, l \in \mathscr{L} = L^2(X, \mu)$. Equation (4.124) is known as the Itô-isometry.

Example 4.107 (The special case of Brownian motion). There are many ways of realizing a Gaussian probability space $(\Omega, \mathcal{F}, \mathbb{P})$. Two candidates for the sample space:

Case 1. Standard Brownian motion process: $\Omega = C(\mathbb{R})$, $\mathcal{F} = \sigma$-algebra generated by cylinder sets, $\mathbb{P} = $ Wiener measure. Set $B_t(\omega) = \omega(t)$, $\forall \omega \in \Omega$; and $\Phi(k) = \int_{\mathbb{R}} k(t) \, dB_t$, $\forall k \in L^2(\Omega, \mathbb{P})$. Case 2. The Gelfand triples:

$\mathcal{S} \hookrightarrow L^2(\mathbb{R}) \hookrightarrow \mathcal{S}'$ (see, e.g., [Trè67, Trè06]), where

(1) $\mathcal{S} = $ the Schwartz space of test functions;
(2) $\mathcal{S}' = $ the space of tempered distributions.

Set $\Omega = \mathcal{S}'$, $\mathcal{F} = \sigma$-algebra generated by cylinder sets of $\mathcal{S}'$, and define

$$\Phi(k) := \widehat{k}(\omega) = \langle k, \omega \rangle, \quad k \in L^2(\mathbb{R}), \ \omega \in \mathcal{S}'.$$

Note Φ is defined by extending the duality $\mathcal{S} \longleftrightarrow \mathcal{S}'$ to $L^2(\mathbb{R})$. The probability measure $\mathbb{P}$ is defined from

$$\mathbb{E}(e^{i\langle k, \cdot \rangle}) = \int_{\mathcal{S}'} e^{i\widehat{k}(\omega)} \, d\mathbb{P}(\omega) = e^{-\frac{1}{2}\|k\|_{L^2(\mathbb{R})}^2},$$

by Minlos' theorem [Hid80, AØ15, Ok08, OR00, Ban97].

Definition 4.108. Let $\mathcal{D} \subset L^2(\Omega, \mathcal{F}, \mathbb{P})$ be the dense subspace spanned by functions F, where $F \in \mathcal{D}$ if and only if $\exists n \in \mathbb{N}$, $\exists h_1, \ldots, h_n \in \mathscr{L}$, and

$p \in \mathbb{R}[x_1, \ldots, x_n]$ = the polynomial ring, such that

$$F = p(\Phi(h_1), \ldots, \Phi(h_n)) : \Omega \longrightarrow \mathbb{R}.$$

(See the diagram below.) The case of $n = 0$ corresponds to the constant function $\mathbb{1}$ on Ω. Note that $\Phi(h_i) \in L^2(\Omega, \mathbb{P})$.

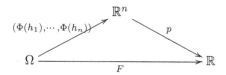

Lemma 4.109. *The polynomial fields $\mathscr{D}$ in Definition 4.108 form a dense subspace in $L^2(\Omega, \mathbb{P})$.*

Proof. The easiest argument below takes advantage of the isometric isomorphism of $L^2(\Omega, \mathbb{P})$ with the symmetric Fock space

$$\Gamma_{sym}(\mathscr{L}) = \underbrace{\mathscr{H}_0}_{\text{1 dim}} \oplus \sum_{n=1}^{\infty} \underbrace{(\mathscr{L} \otimes \cdots \otimes \mathscr{L})}_{n\text{-fold symmetric}}.$$

For $k_i \in \mathscr{L}$, $i = 1, 2$, there is a unique vector $e^{k_i} \in \Gamma_{sym}(\mathscr{L})$ such that

$$\left\langle e^{k_1}, e^{k_2} \right\rangle_{\Gamma_{sym}(\mathscr{L})} = \sum_{n=0}^{\infty} \frac{\langle k_1, k_2 \rangle^n}{n!} = e^{\langle k_1, k_2 \rangle_{\mathscr{L}}}.$$

Moreover,

$$\Gamma_{sym}(\mathscr{L}) \ni e^k \xrightarrow{Z_0} e^{\Phi(k) - \frac{1}{2}\|k\|_{\mathscr{L}}^2} \in L^2(\Omega, \mathbb{P})$$

extends by linearity and closure to a unitary isomorphism $\Gamma_{sym}(\mathscr{L}) \xrightarrow{Z} L^2(\Omega, \mathbb{P})$, mapping onto $L^2(\Omega, \mathbb{P})$ (also see (6.1) in Theorem 6.1.) Hence $\mathscr{D}$ is dense in $L^2(\Omega, \mathbb{P})$, as $span\{e^k \mid k \in \mathscr{L}\}$ is dense in $\Gamma_{sym}(\mathscr{L})$. $\quad\square$

Lemma 4.110. *Let $\mathscr{L}$ be a real Hilbert space, and let $(\Omega, \mathcal{F}, \mathbb{P}, \Phi)$ be an associated Gaussian field. For $n \in \mathbb{N}$, let $\{h_1, \ldots, h_n\}$ be a system of linearly independent vectors in $\mathscr{L}$. Then, for polynomials $p \in \mathbb{R}[x_1, \ldots, x_n]$, the following two conditions are equivalent:*

$$p(\Phi(h_1), \ldots, \Phi(h_n)) = 0 \quad a.e. \text{ on } \Omega \text{ w.r.t } \mathbb{P}; \text{ and} \qquad (4.125)$$

$$p(x_1, \ldots, x_n) \equiv 0, \quad \forall(x_1, \ldots, x_n) \in \mathbb{R}^n. \qquad (4.126)$$

Proof. Let $G_n = (\langle h_i, h_j \rangle)_{i,j=1}^n$ be the Gramian matrix. We have $\det G_n \neq 0$. Let $g^{(G_n)}(x_1, \ldots, x_n)$ be the corresponding Gaussian density;

see (4.119), and Figure 4.8. Then the following are equivalent:

(i) Equation (4.125) holds;

(ii) $p(\Phi(h_1), \ldots, \Phi(h_n)) = 0$ in $L^2(\Omega, \mathcal{F}, \mathbb{P})$;

(iii) $\mathbb{E}\left(|p(\Phi(h_1), \ldots, \Phi(h_n))|^2\right) = \int_{\mathbb{R}^n} |p(x)|^2 g^{(G_n)}(x)\, dx = 0$;

(iv) $p(x) = 0$ a.e. x w.r.t. the Lebesgue measure in $\mathbb{R}^n$;

(v) $p(x) = 0$, $\forall x \in \mathbb{R}^n$; i.e., (4.126) holds. $\qquad\square$

4.5 Equivalence of pairs of Gaussian processes

Consider a pair of Gaussian processes; a natural question is that of deciding when they are equivalent, as opposed to the case when they are mutually singular. While a complete discussion of details is beyond the scope of our book, as with to draw attention to the elegant book [Jr68] by *Ole Groth Jrsboe*. The approach there is especially relevant to our presetting emphasis on *reproducing kernel Hilbert spaces* (RKHSs). Starting with a given pair of Gaussian processes, we then pass to the respective covariance kernels. Note that each of the two processes, say X_1 and X_2 will be governed by respective distributions, and therefore by a pair of probability measures, say $\mathbb{P}_1$ and $\mathbb{P}_2$. We then get a pair of associated reproducing kernel Hilbert spaces RKHS, $\mathscr{H}(K_1)$ and $\mathscr{H}(K_2)$, one for each of the Gaussian processes. With the choice of a pair of corresponding orthonormal bases (ONBs) for $\mathscr{H}(K_1)$ and $\mathscr{H}(K_2)$, Jrsboe is then able to give a precise condition which decides when these two probability measures are equivalent vs mutually relative singular. A key ingredient in the proof is Kakutani's dichotomy result, which asserts that two infinite product measures are either equivalent or pairwise mutually singular.

Theorem 4.111 (Jørsboe). *Let $(\Omega, \mathcal{F})$ be a measurable space, and let $\mathbb{P}_1$ and $\mathbb{P}_2$ be probability measures on $(\Omega, \mathcal{F})$ such that $\{X(t) : t \in T\}$ are Gaussian processes with means $m_1(t)$ and 0, and covariance functions K_1 and K_2, respectively. Denote by $\mathscr{H}(K_i)$ the RKHS of K_i, $i = 1, 2$. Then the known condition for the equivalence of $\mathbb{P}_1$ and $\mathbb{P}_2$ can be stated in terms of the reproducing kernels; namely, they are equivalent if and only if*

(i) *$m_1(\cdot) \in \mathscr{H}(K_2)$; and*

(ii) *K_1 has a representation of the form*

$$K_1(s,t) = \sum_{k=1}^{\infty} \lambda_k e_k(s) e_k(t)$$

with $\sum_{k=1}^{\infty} (1 - \lambda_k)^2 < \infty$ and $\lambda_k > 0$ *for every* k, *where* $\{e_k\}$ *is an orthonormal basis in* $\mathscr{H}(K_2)$.

4.6 Guide to the literature

While we have cited some paper from the Bibliography in our discussion inside the chapter, readers not familiar with the ideas involved may wish to consult the following sources: [AAH69, AH84, AJ12, AJ15, AJS14, AKL16, Aro50, Aro61, AS57, AW73, Bog98, BØSW04, Gro70, Hid71, Hid80, Hid93, Hid03, Hid07, HKPS13, Hud14, Jan97, JT18b, Kat19, Mal78, Sti74, App09]. Also see [Itô06, Itô04, Doo96, Doo89, Doo75, Ĺ92, Ĺ69, KU87, Kol83].

Chapter 5

Infinite-Dimensional Stochastic Analysis: White Noise Analysis and Generalized Itô Calculus

"At that time, few mathematicians regarded probability theory as an authentic mathematical field, in the same strict sense that they regarded differential and integral calculus. With clear definition of real numbers formulated at the end of the 19th century, differential and integral calculus had developed into an authentic mathematical system. When I was a student, there were few researchers in probability; among the few were Kolmogorov of Russia, and Paul Levy of France." Kiyoshi Itô. In Mac Tutor. https://mathshistory.st-andrews.ac.uk/Biographies/Ito/.

"Randomization is too important to be left to chance." J. D. Petruccelli.

While the subject of infinite-dimensional variational calculus, Wiener processes, stochastic differential equations (SDE) is vast, with many and varied subdisciplines, and with many and diverse applications, for our present purposes we have made choices as follows, each motivated by the aim of the book: We shall stress direct connections between Malliavin calculus on the one hand, and the theory of representations of the CCRs on the other. Each one of the two topics throws light on the other, and helps us gain insight. The link between the two will be made precise via our analysis of intertwining operators; see Chapter 7 below. Useful supplement treatments in the literature: [Doo75, Nua06].

In the chapter below, we discuss in more detail notions from infinite-dimensional analysis, with an emphasis on the Gaussian processes, Gaussian fields, Itô calculus, Malliavin derivatives, as they arise in stochastic differential equations, and stochastic PDEs.

5.1 The Malliavin derivatives

> *"The mathematical theory, now known as Malliavin calculus, was first introduced by Paul Malliavin ... as an infinite-dimensional integration by parts technique. The purpose of this calculus was to prove the results about the smoothness of densities of solutions of stochastic differential equations driven by Brownian motion."*
> Giulia Di Nunno, Bernt ksendal, and Frank Proske: *Malliavin Calculus for Lévy Processes with Applications to Finance*; (2008) Springer Science & Business Media.

As mentioned in Chapter 3, the *Malliavin derivative* is the key ingredient in what goes by the name, Malliavin calculus. It is a mathematical framework for *infinite-dimensional analysis*. It is motivated in part by calculus of variations from deterministic functions, but extended to calculus for stochastic processes. Below we shall present Malliavin calculus as part of the study of *representations of the CCRs*.

Below we give an application of the closability criterion for linear operators T between different Hilbert spaces $\mathscr{H}_1$ and $\mathscr{H}_2$, but having dense domain in the first Hilbert space. In this application, we shall take for T to be the so called Malliavin derivative. The setting for it is that of the Wiener process. For the Hilbert space $\mathscr{H}_1$ we shall take the L^2-space, $L^2(\Omega, \mathbb{P})$ where $\mathbb{P}$ is generalized Wiener measure. Below we shall outline the basics of the Malliavin derivative, and we shall specify the two Hilbert spaces corresponding to the setting of Theorem 1.51. We also stress that the literature on Malliavin calculus and its applications is vast, see, e.g., [BØSW04, AØ15, Mal78, Nua06, Bel06].

Settings. It will be convenient for us to work with the *real* Hilbert spaces.

Let $(\Omega, \mathcal{F}, \mathbb{P}, \Phi)$ be as specified in Definition 4.103, i.e., we consider the Gaussian field Φ. Fix a *real* Hilbert space $\mathscr{L}$ with $\dim \mathscr{L} = \aleph_0$. Set $\mathscr{H}_1 = L^2(\Omega, \mathbb{P})$, and $\mathscr{H}_2 = L^2(\Omega \to \mathscr{L}, \mathbb{P}) = L^2(\Omega, \mathbb{P}) \otimes \mathscr{L}$, i.e., vector valued random variables.

For $\mathscr{H}_1$, the inner product $\langle \cdot, \cdot \rangle_{\mathscr{H}_1}$ is

$$\langle F, G \rangle_{\mathscr{H}_1} = \int_\Omega FG \, d\mathbb{P} = \mathbb{E}(FG); \tag{5.1}$$

where $\mathbb{E}(\cdots) = \int_\Omega (\cdots) \, d\mathbb{P}$ is the mean or expectation functional.

On $\mathscr{H}_2$, we have the tensor product inner product: If $F_i \in \mathscr{H}_1$, $k_i \in \mathscr{L}$, $i = 1, 2$, then

$$\langle F_1 \otimes k_1, F_2 \otimes k_2 \rangle_{\mathscr{H}_2} = \langle F_1, F_2 \rangle_{\mathscr{H}_1} \langle k_1, k_2 \rangle_{\mathscr{L}}$$
$$= \mathbb{E}\left(F_1 F_2\right) \langle k_1, k_2 \rangle_{\mathscr{L}}. \tag{5.2}$$

Equivalently, if $\psi_i : \Omega \longrightarrow \mathscr{L}$, $i = 1, 2$, are measurable functions on Ω, we set

$$\langle \psi_1, \psi_2 \rangle_{\mathscr{H}_2} = \int_{\Omega} \langle \psi_1(\omega), \psi_2(\omega) \rangle_{\mathscr{L}} \, d\mathbb{P}(\omega); \tag{5.3}$$

where it is assumed that

$$\int_{\Omega} \|\psi_i(\omega)\|_{\mathscr{L}}^2 \, d\mathbb{P}(\omega) < \infty, \quad i = 1, 2. \tag{5.4}$$

Remark 5.1. In the special case of standard Brownian motion, we have $\mathscr{L} = L^2(0, \infty)$, and set $\Phi(h) = \int_0^\infty h(t) \, d\Phi_t$ (= the Itô-integral), for all $h \in \mathscr{L}$. Recall we then have

$$\mathbb{E}\left(|\Phi(h)|^2\right) = \int_0^\infty |h(t)|^2 \, dt, \tag{5.5}$$

or equivalently (the Itô-isometry),

$$\|\Phi(h)\|_{L^2(\Omega, \mathbb{P})} = \|h\|_{\mathscr{L}}, \quad \forall h \in \mathscr{L}. \tag{5.6}$$

The consideration above also works in the context of general Gaussian fields; see Section 4.4.

Definition 5.2. Let $\mathscr{D}$ be the dense subspace in $\mathscr{H}_1 = L^2(\Omega, \mathbb{P})$ as in Definition 4.108. The operator $T : \mathscr{H}_1 \longrightarrow \mathscr{H}_2$ (= Malliavin derivative) with $dom(T) = \mathscr{D}$ is specified as follows:

For $F \in \mathscr{D}$, i.e., $\exists n \in \mathbb{N}$, $p(x_1, \ldots, x_n)$ a polynomial in n real variables, and $h_1, h_2, \ldots, h_n \in \mathscr{L}$, where

$$F = p(\Phi(h_1), \ldots, \Phi(h_n)) \in L^2(\Omega, \mathbb{P}). \tag{5.7}$$

Set

$$T(F) = \sum_{j=1}^{n} \left(\frac{\partial}{\partial x_j} p\right) (\Phi(h_1), \ldots, \Phi(h_n)) \otimes h_j \in \mathscr{H}_2. \tag{5.8}$$

In the following two remarks we outline the argument for why the expression for $T(F)$ in (5.8) is independent of the chosen representation (5.7) for the particular F. Recall that F is in the domain $\mathscr{D}$ of T. Without some careful justification, it is not even clear that T, as given, defines a linear operator on its dense domain $\mathscr{D}$. The key steps in the argument to follow will be the result (5.12) in Theorem 5.8 below, and the discussion to follow.

There is an alternative argument, based instead on Corollary 2.7; see also Section 6.4 below.

Remark 5.3. It is non-trivial that the formula in (5.8) defines a linear operator. Reason: On the LHS in (5.8), the representation of F from (5.7) is not unique. So we must show that $p(\Phi(h_1), \ldots, \Phi(h_n)) = 0 \implies \text{RHS}_{(5.8)} = 0$ as well. (The dual pair analysis below (see Definition 5.6) is good for this purpose.)

Suppose $F \in \mathscr{D}$ has two representations corresponding to systems of vectors $h_1, \ldots, h_n \in \mathscr{L}$, and $k_1, \ldots, k_m \in \mathscr{L}$, with polynomials $p \in \mathbb{R}[x_1, \ldots, x_n]$, and $q \in \mathbb{R}[x_1, \ldots, x_m]$, where

$$F = p\left(\Phi(h_1), \ldots, \Phi(h_n)\right) = q\left(\Phi(k_1), \ldots, \Phi(k_m)\right). \tag{5.9}$$

We must then verify the identity:

$$\sum_{i=1}^{n} \frac{\partial p}{\partial x_i}(\Phi(h_1), \ldots, \Phi(h_n)) \otimes h_i = \sum_{i=1}^{m} \frac{\partial q}{\partial x_i}(\Phi(k_1), \ldots, \Phi(k_m)) \otimes k_i. \tag{5.10}$$

The significance of the next result is the implication (5.9) $\implies$ (5.10), valid for all choices of representations of the same $F \in \mathscr{D}$. The conclusion from (5.12) in Theorem 5.8 is that the following holds for all $l \in \mathscr{L}$:

$$\mathbb{E}\left(\langle \text{LHS}_{(5.10)}, l \rangle\right) = \mathbb{E}\left(\langle \text{RHS}_{(5.10)}, l \rangle\right) = \mathbb{E}(F\Phi(l)).$$

Moreover, with a refinement of the argument, we arrive at the identity

$$\langle \text{LHS}_{(5.10)} - \text{RHS}_{(5.10)}, G \otimes l \rangle_{\mathscr{H}_2} = 0,$$

valid for all $G \in \mathscr{D}$, and all $l \in \mathscr{L}$.

But $span\left\{G \otimes l \mid G \in \mathscr{D}, l \in \mathscr{L}\right\}$ is dense in $\mathscr{H}_2 \left(= L^2\left(\mathbb{P}\right) \otimes \mathscr{L}\right)$ w.r.t. the tensor-Hilbert norm in $\mathscr{H}_2$ (see (5.2)); and we get the desired identity (5.10) for any two representations of F.

Remark 5.4. An easy case where (5.9) $\implies$ (5.10) can be verified "by hand":

Let $F = \Phi(h)^2$ with $h \in \mathscr{L} \backslash \{0\}$ fixed. We can then pick the two systems $\{h\}$ and $\{h, h\}$ with $p(x) = x^2$, and $q(x_1, x_2) = x_1 x_2$. A direct calculus argument shows that $\text{LHS}_{(5.10)} = \text{RHS}_{(5.10)} = 2\Phi(h) \otimes h \in \mathscr{H}_2$.

We now resume the argument for the general case.

Definition 5.5 (symmetric pair). For $i = 1, 2$, let $\mathscr{H}_i$ be two Hilbert spaces, and suppose $\mathscr{D}_i \subset \mathscr{H}_i$ are given dense subspaces.

We say that a pair of operators (S, T) forms a *symmetric pair* if $dom\,(T) = \mathscr{D}_1$, and $dom\,(S) = \mathscr{D}_2$; and moreover,

$$\langle Tu, v \rangle_{\mathscr{H}_2} = \langle u, Sv \rangle_{\mathscr{H}_1} \tag{5.11}$$

holds for $\forall u \in \mathscr{D}_1$, $\forall v \in \mathscr{D}_2$. (Also see Definition 1.60.)

It is immediate that (5.11) may be rewritten in the form of containment of graphs:

$$T \subset S^*, \quad S \subset T^*.$$

In that case, both S and T are *closable*. We say that a symmetric pair is *maximal* if $\overline{T} = S^*$ and $\overline{S} = T^*$.

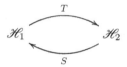

We will establish the following two assertions:

(i) Indeed T from Definition 5.2 is a well-defined linear operator from $\mathscr{H}_1$ to $\mathscr{H}_2$.

(ii) Moreover, (S, T) is a maximal symmetric pair (see Definitions 5.5, 5.6).

Definition 5.6. Let $\mathscr{H}_1 \xrightarrow{T} \mathscr{H}_2$ be the Malliavin derivative with $\mathscr{D}_1 = dom\,(T)$, see Definition 5.2. Set $\mathscr{D}_2 = \mathscr{D}_1 \otimes \mathscr{L} = $ algebraic tensor product, and on $dom\,(S) = \mathscr{D}_2$, set

$$S(F \otimes k) = -\langle T(F), k \rangle + M_{\Phi(k)} F, \quad \forall F \otimes k \in \mathscr{D}_2,$$

where $M_{\Phi(k)} = $ the operator of multiplication by $\Phi(k)$.

Note that both operators S and T are linear and well defined on their respective dense domains, $\mathscr{D}_i \subset \mathscr{H}_i$, $i = 1, 2$. For density, see Lemma 4.109.

It is a "modern version" of ideas in the literature on analysis of Gaussian processes; but we are adding to it, giving it a twist in the direction of multivariable operator theory, representation theory, and especially to representations of infinite-dimensional algebras on generators and relations. Moreover our results apply to more general Gaussian processes than covered so far.

Lemma 5.7. *Let (S, T) be the pair of operators specified above in Definition 5.6. Then it is a symmetric pair, i.e.,*

$$\langle Tu, v \rangle_{\mathscr{H}_2} = \langle u, Sv \rangle_{\mathscr{H}_1}, \quad \forall u \in \mathscr{D}_1, \forall v \in \mathscr{D}_2.$$

Equivalently,

$$\langle T(F), G \otimes k \rangle_{\mathscr{H}_2} = \langle F, S(G \otimes k) \rangle_{\mathscr{H}_1}, \quad \forall F, G \in \mathscr{D}, \forall k \in \mathscr{L}.$$

In particular, we have $S \subset T^*$, and $T \subset S^*$ (containment of graphs.) Moreover, the two operators $S^*\overline{S}$ and $T^*\overline{T}$ are selfadjoint. (For the last conclusion in the lemma, see Theorem 1.51.)

Theorem 5.8. *Let $T : \mathscr{H}_1 \longrightarrow \mathscr{H}_2$ be the Malliavin derivative, i.e., T is an unbounded closable operator with dense domain $\mathscr{D}$ consisting of the span of all the functions F from (5.7). Then, for all $F \in dom(T)$, and $k \in \mathscr{L}$, we have*

$$\mathbb{E}(\langle T(F), k \rangle_{\mathscr{L}}) = \mathbb{E}(F\Phi(k)). \tag{5.12}$$

Proof. We shall prove (5.12) in several steps. Once (5.12) is established, then there is a recursive argument which yields a dense subspace in $\mathscr{H}_2$, contained in $dom(T^*)$; and so T is closable.

Moreover, formula (5.12) yields directly the evaluation of $T^* : \mathscr{H}_2 \longrightarrow \mathscr{H}_1$ as follows: If $k \in \mathscr{L}$, set $\mathbb{1} \otimes k \in \mathscr{H}_2$ where $\mathbb{1}$ denotes the constant function "one" on Ω. We get

$$T^*(\mathbb{1} \otimes k) = \Phi(k) = \int_0^\infty k(t)\, d\Phi_t \, (= \text{the Itô-integral.}) \tag{5.13}$$

The same argument works for any Gaussian field; see Definition 4.103. We refer to the literature [BØSW04, AØ15] for details.

The proof of (5.12) works for any Gaussian process $\mathscr{L} \ni k \longrightarrow \Phi(k)$ indexed by an arbitrary Hilbert space $\mathscr{L}$ with the inner product $\langle k, l \rangle_{\mathscr{L}}$ as the covariance kernel.

Formula (5.12) will be established as follows: Let F and $T(F)$ be as in (5.7)–(5.8).

Step 1. For every $n \in \mathbb{N}$, the polynomial ring $\mathbb{R}[x_1, x_2, \ldots, x_n]$ is invariant under matrix substitution $y = Mx$, where M is an $n \times n$ matrix over $\mathbb{R}$.

Step 2. Hence, in considering (5.12) for $\{h_i\}_{i=1}^{n} \subset \mathcal{L}$, $h_1 = k$, we may diagonalize the $n \times n$ Gramian matrix $(\langle h_i, h_j \rangle)_{i,j=1}^{n}$; thus without loss of generality, we may assume that the system $\{h_i\}_{i=1}^{n}$ is orthogonal and normalized, i.e., that

$$\langle h_i, h_j \rangle = \delta_{ij}, \ \forall i, j \in \{1, \ldots, n\}, \tag{5.14}$$

and we may take $k = h_1$ in $\mathcal{L}$.

Step 3. With this simplification, we now compute the LHS in (5.12). We note that the joint distribution of $\{\Phi(h_i)\}_{i=1}^{n}$ is thus the standard Gaussian kernel in $\mathbb{R}^n$, i.e.,

$$g_n(x) = (2\pi)^{-n/2} e^{-\frac{1}{2} \sum_{i=1}^{n} x_i^2}, \tag{5.15}$$

with $x = (x_1, \ldots, x_n) \in \mathbb{R}^n$. We have

$$x_1 g_n(x) = -\frac{\partial}{\partial x_1} g_n(x) \tag{5.16}$$

by calculus.

Step 4. A direct computation yields

$$
\begin{aligned}
\text{LHS}_{(5.12)} \quad &= \quad \mathbb{E}(\langle T(F), h_1 \rangle_{\mathcal{L}}) \\
&\overset{\text{by (5.14)}}{=} \quad \mathbb{E}(\frac{\partial p}{\partial x_1}(\Phi(h_1), \ldots \Phi(h_n))) \\
&\overset{\text{by (5.15)}}{=} \quad \int_{\mathbb{R}^n} \frac{\partial p}{\partial x_1}(x_1, \ldots, x_n) g_n(x_1, \ldots, x_n) dx_1 \cdots dx_n \\
&\overset{\text{int. by parts}}{=} \quad -\int_{\mathbb{R}^n} p(x_1, \ldots, x_n) \frac{\partial g_n}{\partial x_1}(x_1, \ldots, x_n) dx_1 \cdots dx_n \\
&\overset{\text{by (5.16)}}{=} \quad \int_{\mathbb{R}^n} x_1 p(x_1, \ldots, x_n) g_n(x_1, \ldots, x_n) dx_1 \ldots dx_n \\
&\overset{\text{by (5.14)}}{=} \quad \mathbb{E}(\Phi(h_1) p(\Phi(h_1), \ldots, \Phi(h_n))) \\
&= \quad \mathbb{E}(\Phi(h_1) F) = \text{RHS}_{(5.12)},
\end{aligned}
$$

which is the desired conclusion (5.12). $\quad\square$

Corollary 5.9. *Let $\mathcal{H}_1$, $\mathcal{H}_2$, and $\mathcal{H}_1 \overset{T}{\longrightarrow} \mathcal{H}_2$ be as in Theorem 5.8, i.e., T is the Malliavin derivative. Then, for all $h, k \in \mathcal{L} = L^2(0, \infty)$, we have* *for the closure $\overline{T}$ of T the following:*

$$\overline{T}(e^{\Phi(h)}) = e^{\Phi(h)} \otimes h, \quad and \tag{5.17}$$

$$\mathbb{E}(\langle \overline{T}(e^{\Phi(h)}), k \rangle_{\mathscr{L}}) = e^{\frac{1}{2}\|h\|^2_{\mathscr{L}}} \langle h, k \rangle_{\mathscr{L}}. \tag{5.18}$$

Here $\overline{T}$ denotes the graph-closure of T.

Moreover,

$$T^*\overline{T}(e^{\Phi(k)}) = \left(\Phi(k) - \|k\|^2_{\mathscr{L}} \right) e^{\Phi(k)}. \tag{5.19}$$

Proof. Equations (5.17)–(5.18) follow immediately from (5.12) and a polynomial approximation to

$$e^x = \lim_{n\to\infty} \sum_0^n \frac{x^j}{j!}, \quad x \in \mathbb{R};$$

see (5.7). In particular, $e^{\Phi(h)} \in dom\left(\overline{T}\right)$, and $\overline{T}\left(e^{\Phi(h)}\right)$ is well defined.

For (5.19), we use the facts for the Gaussians:

$$\mathbb{E}(e^{\Phi(k)}) = e^{\frac{1}{2}\|k\|^2}, \quad and$$

$$\mathbb{E}(\Phi(k)\, e^{\Phi(k)}) = \|k\|^2\, e^{\frac{1}{2}\|k\|^2}. \qquad \square$$

Example 5.10. Let $F = \Phi(k)^n$, $\|k\| = 1$. We have

$$T\Phi(k)^n = n\Phi(k)^{n-1} \otimes k$$

$$T^*T\Phi(k)^n = -n(n-1)\Phi(k)^{n-2} + n\Phi(k)^n$$

and similarly,

$$\overline{T}e^{\Phi(k)} = e^{\Phi(k)} \otimes k$$

$$T^*\overline{T}e^{\Phi(k)} = e^{\Phi(k)}(\Phi(k) - 1).$$

Let (S, T) be the symmetric pair, we then have the inclusion $\overline{T} \subset S^*$, i.e., containment of the operator graphs, $\mathscr{G}\left(\overline{T}\right) \subset \mathscr{G}(S^*)$. In fact, we have

Corollary 5.11. $\overline{T} = S^*$.

Proof. We will show that $\mathscr{G}(S^*) \ominus \mathscr{G}\left(\overline{T}\right) = 0$, where $\ominus$ stands for the orthogonal complement in the direct sum-inner product of $\mathscr{H}_1 \oplus \mathscr{H}_2$. Recall that $\mathscr{H}_1 = L^2(\Omega, \mathbb{P})$, and $\mathscr{H}_2 = \mathscr{H}_1 \otimes \mathscr{L}$.

Using (5.17), we will prove that if $F \in dom\,(S^*)$, and

$$\left\langle \begin{pmatrix} e^{\Phi(k)} \\ e^{\Phi(k)} \otimes k \end{pmatrix}, \begin{pmatrix} F \\ S^*F \end{pmatrix} \right\rangle = 0, \; \forall k \in \mathscr{L} \Longrightarrow F = 0,$$

which is equivalent to

$$\mathbb{E}\left(e^{\Phi(k)}\left(F + \langle S^*F, k\rangle\right)\right) = 0, \; \forall k \in \mathscr{L}. \tag{5.20}$$

But it is know that for the Gaussian filed, $span\left\{e^{\Phi(k)} \mid k \in \mathscr{L}\right\}$ is dense in $\mathscr{H}_1$, and so (5.20) implies that $F = 0$, which is the desired conclusion.

We can finish the proof of the corollary with an application of Girsanov's theorem, see, e.g., [BØSW04] and [Pri10]. By this result, we have a measurable action τ of $\mathscr{L}$ on $(\Omega, \mathcal{F}, \mathbb{P})$, i.e.,

$$\mathscr{L} \xrightarrow{\tau} Aut\,(\Omega, \mathcal{F})$$

$$\tau_k \circ \tau_l = \tau_{k+l} \quad \text{a.e. on } \Omega, \; \forall k, l \in \mathscr{L} \tag{5.21}$$

(see also Section 6.4 below) such that $\tau_k\,(\mathcal{F}) = \mathcal{F}$ for all $k \in \mathscr{L}$, and

$$\mathbb{P} \circ \tau_k^{-1} \ll \mathbb{P}$$

with

$$\frac{d\mathbb{P} \circ \tau_k^{-1}}{d\mathbb{P}} = e^{-\frac{1}{2}\|k\|_{\mathscr{L}}^2} e^{\Phi(k)}, \quad \text{a.e. on } \Omega. \tag{5.22}$$

Returning to (5.20). An application of (5.22) to (5.20) yields:

$$F\,(\cdot + k) + \langle S^*\,(F)\,(\cdot + k), k\rangle_{\mathscr{L}} = 0 \quad \text{a.e. on } \Omega; \tag{5.23}$$

where we have used "$\cdot + k$" for the action in (5.21). Since τ in (5.21) is an action by measure-automorphisms, (5.23) implies

$$F\,(\cdot) + \langle S^*\,(F)\,(\cdot), k\rangle_{\mathscr{L}} = 0; \tag{5.24}$$

again with $k \in \mathscr{L}$ arbitrary. If $F \neq 0$ in $L^2\,(\Omega, \mathcal{F}, \mathbb{P})$, then the second term in (5.24) would be independent of k which is impossible with $S^*\,(F)\,(\cdot) \neq 0$. But if $S^*\,(F) = 0$, then $F\,(\cdot) = 0$ (in $L^2\,(\Omega, \mathcal{F}, \mathbb{P})$) by (5.24); and so the proof is completed. $\qquad\square$

Remark 5.12. We recall the definition of the domain of the closure $\overline{T}$. The following is a necessary and sufficient condition for an $F \in L^2\,(\Omega, \mathcal{F}, \mathbb{P})$ to be in the domain of $\overline{T}$:

$$F \in dom\left(\overline{T}\right) \iff \exists \text{ a sequence } \{F_n\} \subset \mathscr{D} \text{ such that}$$

$$\lim_{n,m \to \infty} \mathbb{E}\left(|F_n - F_m|^2 + \|T\left(F_n\right) - T\left(F_m\right)\|_{\mathscr{L}}^2\right) = 0. \tag{5.25}$$

When (5.25) holds, we have:

$$\overline{T}\left(F\right) = \lim_{n \to \infty} T\left(F_n\right) \tag{5.26}$$

where the limit on the RHS in (5.26) is in the Hilbert norm of $L^2\left(\Omega, \mathcal{F}, \mathbb{P}\right) \otimes \mathscr{L}$.

Corollary 5.13. *Let $\left(\mathscr{L}, \Omega, \mathcal{F}, \mathbb{P}, \Phi\right)$ be as above, and let T and S be the two operators from Corollary 5.11. Then, for the domain of $\overline{T}$, we have the following:*

For random variables F in $L^2\left(\Omega, \mathcal{F}, \mathbb{P}\right)$, the following two conditions are equivalent:

(i) $F \in dom\left(\overline{T}\right)$;
(ii) $\exists C = C_F < \infty$ *such that*

$$|\mathbb{E}\left(F\, S\left(\psi\right)\right)|^2 \leq C\, \mathbb{E}\left(\|\psi\left(\cdot\right)\|_{\mathscr{L}}^2\right)$$

holds for $\forall \psi \in span\,\{G \otimes k \mid G \in \mathscr{D},\ k \in \mathscr{L}\}$.
Recall

$$S\left(\cdot \otimes k\right) = M_{\Phi(k)} \cdot - \langle T(\cdot), k \rangle_{\mathscr{L}};$$

equivalently,

$$S\left(G \otimes k\right) = \Phi\left(k\right) G - \langle T(G), k \rangle_{\mathscr{L}}$$

for all $G \in \mathscr{D}$, and all $k \in \mathscr{L}$.

Proof. Immediate from the previous corollary.   □

Lemma 5.14. *Let $\{h_i\}_{i=1}^n \subset \mathscr{L}$, and*

$$G^{(n)} = \exp\left(-\frac{1}{2}\sum_{ij} C_{ij}\Phi\left(h_i\right)\Phi\left(h_j\right)\right), \tag{5.27}$$

where

$$C^{-1} = \left(\langle \Phi(h_i), \Phi(h_j)\rangle_{\mathscr{H}_1}\right) = \left(\langle h_i, h_j\rangle_{\mathscr{L}}\right), \tag{5.28}$$

i.e., the Gramian matrix.

Then $-T\left(G^{(n)}\right) = \sum_{i=1}^{n} C_{ij}\Phi\left(h_i\right) \otimes h_i$, *which may be identified with the operator* $\mathscr{L} \to \mathscr{H}_1$,

$$-T\left(G^{(n)}\right)(\cdot) = \left[\Phi\left(h_1\right) \cdots \Phi\left(h_n\right)\right] C \begin{bmatrix} \langle h_1, \cdot \rangle_{\mathscr{L}} \\ \vdots \\ \langle h_n, \cdot \rangle_{\mathscr{L}} \end{bmatrix}. \tag{5.29}$$

Moreover, if $\widetilde{P}_n$ *is the projection from* $\mathscr{H}_1 = L^2\left(\Omega, \mathbb{P}\right)$ *onto the n-dimensional subspace* $\mathscr{K}_n = span\left\{\Phi\left(h_i\right)\right\}$, *then*

$$-T\left(G^{(n)}\right)(k) = \widetilde{P}_n\left(\Phi\left(k\right)\right), \ k \in \mathscr{L}. \tag{5.30}$$

Proof. Assume the system $\{h_i\}_{i=1}^{n} \subset \mathscr{L}$ is linearly independent but not necessarily orthogonal. The projection P_n from $\mathscr{L}$ onto $span\{h_i\}$ is given by

$$\mathscr{L} \ni k \xrightarrow{P_n} \left[h_1 \cdots h_n\right] C \begin{bmatrix} \langle h_1, k \rangle_{\mathscr{L}} \\ \vdots \\ \langle h_n, k \rangle_{\mathscr{L}} \end{bmatrix}. \tag{5.31}$$

The assertion follows from the fact that $\Phi : \mathscr{L} \to \mathscr{H}_1$ is an isometry. In fact, it holds that

$$\Phi \circ P_n = \widetilde{P}_n \circ \Phi. \tag{5.32}$$

In particular,

$$\widetilde{P}_n\left(\Phi\left(k\right)\right) = \Phi\left(P_n k\right)$$
$$= \Phi\left(\text{RHS of } (5.31)\right)$$
$$= \text{LHS of } (5.29),$$

which is the conclusion of (5.29). $\qquad\square$

Corollary 5.15. *For all* $F = F(\Phi(h_1), \dots \Phi(h_n)) \in \mathscr{H}_1$, *and* $H \otimes k \in \mathscr{H}_2$, *we have*

$$\langle T\left(F\right), H \otimes k \rangle_{\mathscr{H}_2} = \langle F, \langle -T(H), k \rangle_{\mathscr{L}} + \Phi(k)H \rangle_{\mathscr{H}_1}, \tag{5.33}$$

which, in turn, yields the adjoint of T, *i.e.,*

$$T^*(H \otimes k) = \langle -T(H), k \rangle_{\mathscr{L}} + \Phi(k)H. \tag{5.34}$$

Proof. Note that

$$\langle T(F), H \otimes k \rangle_{\mathscr{H}_2} = \sum_{i=1}^{n} \left\langle \frac{\partial F}{\partial x_i}, H \right\rangle_{\mathscr{H}_1} \langle h_i, k \rangle_{\mathscr{L}} = \sum_{i=1}^{n} \left\langle F, -\frac{\partial H}{\partial x_i} \right\rangle_{\mathscr{H}_1}$$

$$\times \langle h_i, k \rangle_{\mathscr{L}} + \sum_{i=1}^{n} \left\langle F, -\frac{\partial G^{(n)}}{\partial x_i} H \right\rangle_{\mathscr{H}_1} \langle h_i, k \rangle_{\mathscr{L}}$$

$$= \langle F, \langle -T(H), k \rangle_{\mathscr{L}} \rangle_{\mathscr{H}_1} + \left\langle F, -T(G^{(n)})(k) H \right\rangle_{\mathscr{H}_1}$$

$$= \langle F, \langle -T(H), k \rangle_{\mathscr{L}} \rangle_{\mathscr{H}_1} + \underbrace{\left\langle F, \widetilde{P}_n \Phi(k) H \right\rangle_{\mathscr{H}_1}}_{\text{by Lemma 5.14}}$$

$$= \langle F, \langle -T(H), k \rangle_{\mathscr{L}} \rangle_{\mathscr{H}_1} + \langle F, \Phi(k) H \rangle_{\mathscr{H}_1}$$

$$\left(\text{since } \widetilde{P}_n F = F \right) = \langle F, \langle -T(H), k \rangle_{\mathscr{L}} + \Phi(k) H \rangle_{\mathscr{H}_1},$$

and so (5.33) & (5.34) hold. $\qquad\qquad\qquad\qquad\qquad\qquad\qquad\qquad\square$

5.2   A derivation on the algebra $\mathscr{D}$

The study of unbounded derivations has many applications in mathematical physics; in particular in making precise the time dependence of quantum observables, i.e., the dynamics in the Schrödinger picture; — in more detail, in the problem of constructing dynamics in statistical mechanics. An early application of unbounded derivations (in the commutative case) can be found in the work of Šilov [Šil47]; and the later study of unbounded derivations in non-commutative C^*-algebras is outlined in [BR81]. There is a rich in variety unbounded derivations, because of the role they play in applications to dynamical systems in quantum physics. See, e.g., [BJr82, BJr83, BEJ84, BJ89, BEGJ89b, BEGJ89a].

But previously the theory of unbounded derivations has not yet been applied systematically to stochastic analysis in the sense of Malliavin (see Section 5.1). In the present section, we turn to this. We begin with the following:

Lemma 5.16 (Leibniz–Malliavin). *Let* $\mathscr{H}_1 \xrightarrow{\;T\;} \mathscr{H}_2$ *be the Malliavin derivative from* (5.7)–(5.8). *Then,*

   (i) *dom* $(T) =: \mathscr{D}$, *given by* (5.7), *is an* algebra *of functions on* Ω *under pointwise product, i.e.,* $FG \in \mathscr{D}$, $\forall F, G \in \mathscr{D}$.

   (ii) $\mathscr{H}_2$ *is a* module *over* $\mathscr{D}$ *where* $\mathscr{H}_2 = L^2(\Omega, \mathbb{P}) \otimes \mathscr{L}$ (= *vector valued* L^2-*random variables.*)

(iii) *Moreover,*

$$T(FG) = T(F)G + FT(G), \quad \forall F, G \in \mathscr{D}, \tag{5.35}$$

i.e., T is a module-derivation.

Notation. Equation (5.35) *is called the Leibniz-rule. By the Leibniz, we refer to the traditional rule of Leibniz for the derivative of a product. And the Malliavin derivative is thus an infinite-dimensional extension of Leibniz calculus.*

Proof. To show that $\mathscr{D} \subset \mathscr{H}_1 = L^2(\Omega, \mathbb{P})$ is an algebra under point-wise multiplication, the following trick is useful. It follows from finite-dimensional Hilbert space geometry.

Let F, G be as in Definition 4.108. Then $\exists p, q \in \mathbb{R}[x_1, \ldots, x_n]$, $\{l_i\}_{i=1}^n \subset \mathscr{L}$, such that

$$F = p(\Phi(l_1), \ldots, \Phi(l_n)), \quad \text{and} \quad G = q(\Phi(l_1), \ldots, \Phi(l_n)).$$

That is, the same system $l_1, \ldots, l_n$ may be chosen for the two functions F and G.

For the pointwise product, we have

$$FG = (pq)(\Phi(l_1), \ldots, \Phi(l_n)),$$

i.e., the product in $\mathbb{R}[x_1, \ldots, x_n]$ with substitution of the random variable

$$(\Phi(l_1), \ldots, \Phi(l_n)) : \Omega \longrightarrow \mathbb{R}^n.$$

Equation (5.35) $\Longleftrightarrow \frac{\partial(pq)}{\partial x_i} = \frac{\partial p}{\partial x_i}q + p\frac{\partial q}{\partial x_i}$, which is the usual Leibniz rule applied to polynomials. Note that

$$T(FG) = \sum_{i=1}^n \frac{\partial}{\partial x_i}(pq)(\Phi(l_1), \ldots, \Phi(l_n)) \otimes l_i.$$

$\square$

Remark 5.17. There is an extensive literature on the theory of densely defined unbounded derivations in C^*-algebras. This includes both the cases of abelian and non-abelian *-algebras. And moreover, this study includes both derivations in these algebras, as well as the parallel study of module derivations. So the case of the Malliavin derivative is in fact a special case of this study. Readers interested in details are referred to [Sak98], [BJKR84], [BR79], and [BR81].

Definition 5.18. Let $(\mathscr{L}, \Omega, \mathcal{F}, \mathbb{P}, \Phi)$ be a Gaussian field, and T be the Malliavin derivative with $dom\,(T) = \mathscr{D}$. For all $k \in \mathscr{L}$, set

$$T_k\,(F) := \langle T\,(F)\,, k \rangle\,, \quad F \in \mathscr{D}. \tag{5.36}$$

In particular, let $F = p(\Phi(l_1), \dots, \Phi(l_1))$ be as in (5.7), then

$$T_k(F) = \sum_{i=1}^{n} \frac{\partial p}{\partial x_i}(\Phi(l_1), \dots, \Phi(l_1))\langle l_i, k \rangle.$$

Corollary 5.19. T_k *is a derivative on* $\mathscr{D}$, *i.e.,*

$$T_k\,(FG) = (T_k F)\,G + F\,(T_k G)\,, \quad \forall F, G \in \mathscr{D}, \forall k \in \mathscr{L}. \tag{5.37}$$

Proof. Follows from (5.35). $\qquad\square$

Corollary 5.20. *Let* $(\mathscr{L}, \Omega, \mathcal{F}, \mathbb{P}, \Phi)$ *be a Gaussian field. Fix* $k \in \mathscr{L}$, *and let* T_k *be the Malliavin derivative in the k direction. Then on* $\mathscr{D}$ *we have*

$$T_k + T_k^* = M_{\Phi(k)}, \quad and \tag{5.38}$$

$$[T_k, T_l^*] = \langle k, l \rangle_{\mathscr{L}}\, I_{L^2(\Omega, \mathbb{P})}. \tag{5.39}$$

Proof. For all $F, G \in \mathscr{D}$, we have

$$\mathbb{E}\,(T_k\,(F)\,G) + \mathbb{E}\,(F\,T_k\,(G)) \stackrel{\text{by (5.37)}}{=} \mathbb{E}\,(T_k\,(FG))$$
$$\stackrel{\text{by (5.12)}}{=} \mathbb{E}\,(\Phi\,(k)\,FG)$$

which yields the assertion in (5.38). Equation (5.39) now follows from (5.38) and the fact that $[T_k, T_l] = 0$. $\qquad\square$

Definition 5.21. Let $(\mathscr{L}, \Omega, \mathcal{F}, \mathbb{P}, \Phi)$ be a Gaussian field. For all $k \in \mathscr{L}$, let T_k be Malliavin derivative in the k-direction (see (5.36)). Assume $\mathscr{L}$ is separable, i.e., $\dim \mathscr{L} = \aleph_0$. For every ONB $\{e_i\}_{i=1}^{\infty}$ in $\mathscr{L}$, let

$$N := \sum_i T_{e_i}^* T_{e_i}. \tag{5.40}$$

(N is the CCR number operator. See Section 6.3 below.)

Example 5.22. $N\mathbb{1} = 0$, since $T_{e_i}\mathbb{1} = 0$, $\forall i$. Similarly,

$$N\Phi\,(k) = \Phi\,(k) \tag{5.41}$$

$$N\Phi\,(k)^2 = -2\,\|k\|^2\,\mathbb{1} + 2\Phi\,(k)^2\,, \quad \forall k \in \mathscr{L}. \tag{5.42}$$

To see this, note that

$$\sum_i T^*_{e_i} T_{e_i} \Phi(k) = \sum_i T^*_{e_i} \langle e_i, k \rangle \mathbb{1}$$

$$= \sum_i \Phi(e_i) \langle e_i, k \rangle$$

$$= \Phi\left(\sum_i \langle e_i, k \rangle e_i\right) = \Phi(k),$$

which is (5.41). The verification of (5.42) is similar.

Theorem 5.23. *Let $\{e_i\}$ be an ONB in $\mathscr{L}$, then*

$$T^*\overline{T} = \sum_i T^*_{e_i} T_{e_i} = N. \tag{5.43}$$

Proof. Note the span of $\{e^{\Phi(k)} \mid k \in \mathscr{L}\}$ is dense in $L^2(\Omega, \mathbb{P})$, and both sides of (5.43) agree on $e^{\Phi(k)}$, $k \in \mathscr{L}$. Indeed, by (5.40),

$$T^*\overline{T} e^{\Phi(k)} = N e^{\Phi(k)} = \left(\Phi(k) - \|k\|^2\right) e^{\Phi(k)}.$$

$\square$

Corollary 5.24. *Let $D := T^*\overline{T}$. Specialize to the case of $n = 1$, and consider $F = f(\Phi(k))$, $k \in \mathscr{L}$, $f \in C^\infty(\mathbb{R})$; then*

$$D(F) = -\|k\|^2_{\mathscr{L}} f''(\Phi(k)) + \Phi(k) f'(\Phi(k)). \tag{5.44}$$

Proof. A direct application of the formulas of $\overline{T}$ and T^*. $\square$

Remark 5.25. If $\|k\|_{\mathscr{L}} = 1$ in (5.44), then the RHS in (5.44) is obtained by a substitution of the real valued random variable $\Phi(k)$ into the deterministic function

$$\delta(f) := -\left(\frac{d}{dx}\right)^2 f + x \left(\frac{d}{dx}\right) f. \tag{5.45}$$

Then (5.44) may be rewritten as

$$D(f(\Phi(k))) = \delta(f) \circ \Phi(k), \quad f \in C^\infty(\mathbb{R}). \tag{5.46}$$

Corollary 5.26. *If $\{H_n\}_{n \in \mathbb{N}_0}$, $\mathbb{N}_0 = \{0, 1, 2, \ldots\}$, denotes the Hermite polynomials on $\mathbb{R}$, then we get for $\forall k \in \mathscr{L}$, $\|k\|_{\mathscr{L}} = 1$, the following eigenvalues*

$$D(H_n(\Phi(k))) = n H_n(\Phi(k)). \tag{5.47}$$

Proof. It is well-known that the Hermite polynomials H_n satisfies

$$\delta\left(H_n\right) = n\,H_n, \quad \forall n \in \mathbb{N}_0, \tag{5.48}$$

and so (5.47) follows from a substitution of (5.48) into (5.46).   □

Theorem 5.27. *The spectrum of $T^*\overline{T}$, as an operator in $L^2\left(\Omega, \mathcal{F}, \mathbb{P}\right)$, is as follows*:

$$spec_{L^2(\mathbb{P})}\left(T^*\overline{T}\right) = \mathbb{N}_0 = \{0, 1, 2, \ldots\}.$$

Proof. We saw that the $L^2\left(\mathbb{P}\right)$-representation is unitarily equivalent to the Fock vacuum representation, and π (Fock-number operator) $= T^*\overline{T}$.   □

5.3 Infinite-dimensional Δ and ∇_Φ

The purpose of the present section is to show how our presentation of a calculus for Gaussian field, outlined in the first half of Chapter 5, yield streamlined presentations of the two operators in infinite-dimensional calculus, the Laplacian and the gradients. (Our present focus to stochastic analysis is that which is based on representation theory.) Note that variants of these operators often go by the name "*stochastic Laplacian*" and "*stochastic gradients*."

Corollary 5.28. *Let $(\mathcal{L}, \Omega, \mathcal{F}, \mathbb{P}, \Phi)$ be a Gaussian field, and let T be the Malliavin derivative, $L^2(\Omega, \mathbb{P}) \overset{T}{\longrightarrow} L^2\left(\Omega, \mathbb{P}\right) \otimes \mathcal{L}$. Then, for all $F = p(\Phi(h_1), \ldots, \Phi(h_n)) \in \mathcal{D}$ (see Definition 5.2), we have*

$$T^*T\left(F\right) = \underbrace{-\sum_{i=1}^{n} \frac{\partial^2 p}{\partial x_i}(\Phi(h_1), \ldots, \Phi(h_n))}_{\Delta F}$$

$$\underbrace{+\sum_{i=1}^{n} \Phi\left(h_i\right) \frac{\partial p}{\partial x_i}(\Phi(h_1), \ldots, \Phi(h_n)),}_{\nabla_\Phi F}$$

which is abbreviated

$$T^*T = -\Delta + \nabla_\Phi. \tag{5.49}$$

(For the general theory of infinite-dimensional Laplacians, see, e.g., [Hid03].)

Proof. (Sketch) We may assume the system $\{h_i\}_{i=1}^n \subset \mathcal{L}$ is orthonormal, i.e., $\langle h_i, h_j \rangle = \delta_{ij}$. Hence, for F $F = p(\Phi(h_1), \ldots, \Phi(h_n)) \in \mathcal{D}$, we have

$$TF = \sum_{i=1}^n \frac{\partial p}{\partial x_i}(\Phi(h_1), \ldots, \Phi(h_n)) \otimes h_i, \quad \text{and}$$

$$T^*T(F) = -\sum_{i=1}^n \frac{\partial^2 p}{\partial x_i^2}(\Phi(h_1), \ldots, \Phi(h_n))$$

$$+ \sum_{i=1}^n \Phi(h_i) \frac{\partial p}{\partial x_i}(\Phi(h_1), \ldots, \Phi(h_n))$$

which is the assertion. For details, see the proof of Theorem 5.8. $\square$

Definition 5.29. Let $(\mathcal{L}, \Omega, \mathcal{F}, \mathbb{P}, \Phi)$ be a Gaussian field. On the dense domain $\mathcal{D} \subset L^2(\Omega, \mathbb{P})$, we define the Φ-gradient by

$$\nabla_\Phi F = \sum_{i=1}^n \Phi(h_i) \frac{\partial p}{\partial x_i}(\Phi(h_1), \ldots, \Phi(h_n)), \tag{5.50}$$

for all $F = p(\Phi(h_1), \ldots, \Phi(h_n)) \in \mathcal{D}$. (Note that ∇_Φ is an unbounded operator in $L^2(\Omega, \mathbb{P})$, and $dom(\nabla_\Phi) = \mathcal{D}$.)

Lemma 5.30. *Let ∇_Φ be the Φ-gradient from Definition 5.29. The adjoint operator ∇_Φ^*, i.e., the Φ-divergence, is given as follows:*

$$\nabla_\Phi^*(G) = \left(\sum_{i=1}^n \Phi(h_i)^2 - n \right) G - \nabla_\Phi(G), \quad \forall G \in \mathcal{D}. \tag{5.51}$$

Proof. Fix $F, G \in \mathcal{D}$ as in Definition 5.2. Then $\exists n \in \mathbb{N}$, $p, q \in \mathbb{R}[x_1, \ldots, x_n]$, and $\{h_i\}_{i=1}^n \subset \mathcal{L}$, such that

$$F = p(\Phi(h_1), \ldots, \Phi(h_n))$$

$$G = q(\Phi(h_1), \ldots, \Phi(h_n)).$$

Further assume that $\langle h_i, h_j \rangle = \delta_{ij}$.

In the calculation below, we use the following notation: $x = (x_1, \ldots, x_n) \in \mathbb{R}^n$, $dx = dx_1 \ldots dx_n = $ Lebesgue measure, and $g_n = g^{G_n} = $ standard Gaussian distribution in $\mathbb{R}^n$, see (5.15).

Then, we have

$$\mathbb{E}((\nabla_\Phi F)G)$$

$$= \sum_{i=1}^n \mathbb{E}\left(\Phi(h_i)\frac{\partial p}{\partial x_i}(\Phi(h_1),\dots,\Phi(h_n))q(\Phi(h_1),\dots,\Phi(h_n))\right)$$

$$= \sum_{i=1}^n \int_{\mathbb{R}^n} x_i \frac{\partial p}{\partial x_i}(x)q(x)g_n(x)dx$$

$$= -\sum_{i=1}^n \int_{\mathbb{R}^n} p(x)\frac{\partial}{\partial x_i}(x_i q(x)g_n(x))dx$$

$$= -\sum_{i=1}^n \int_{\mathbb{R}^n} p(x)(q(x) + x_i\frac{\partial q}{\partial x_i}(x) - q(x)x_i^2)g_n(x)dx$$

$$\left(\frac{\partial g_n}{\partial x_i} = -x_i g_n\right) = \sum_{i=1}^n \mathbb{E}\left(FG\Phi(h_i)^2\right) - n\mathbb{E}(FG) - \mathbb{E}(F\nabla_\Phi G)$$

$$= \mathbb{E}\left(FG\left(\sum_{i=1}^n \Phi(h_i)^2 - n\right)\right) - \mathbb{E}(F\nabla_\Phi G),$$

which is the desired conclusion in (5.51). $\qquad\square$

Remark 5.31. Note T_k^* is *not* a derivation. In fact, we have

$$T_k^*(FG) = T_k^*(F)G + FT_k^*(G) - \Phi(k)FG,$$

for all $F, G \in \mathscr{D}$, and all $k \in \mathscr{L}$.

However, the divergence operator ∇_Φ does satisfy the Leibniz rule, i.e.,

$$\nabla_\Phi(FG) = (\nabla_\Phi F)G + F(\nabla_\Phi G), \quad \forall F, G \in \mathscr{D}.$$

5.4 Guide to the literature

While we have cited some paper from the Bibliography in our discussion inside the chapter, readers not familiar with the ideas involved may wish to consult the following sources: [AH84, Mal78, AKL16, AØ15, BØSW04, Hid03, Nua06, Sti74].

Chapter 6

Representations of the CCRs Realized as Gaussian Fields and Malliavin Derivatives

> "... *statistics is not a subset of mathematics, and calls for skills and judgment that are not exclusively mathematical. On the other hand, there is a large intersection between the two disciplines, statistical theory is serious mathematics, and most of the fundamental advances, even in applied statistics, have been made by mathematicians like R. A. Fisher.*" Sir John Kingman, interview in the EMS Newsletter, March 2002.

While the subject of infinite-dimensional calculus via Malliavin derivatives is vast, with many and varied subdisciplines, and with many and diverse applications, for our present purposes we have made choices as follows, each motivated by the aim of the book: our focus will be analysis and representation theory for the CCRs, with an emphasis on the case of the Fock-state (ground state) model, as well as wider classes of representations of CCRs, and their realization in Malliavin's calculus of variation. Useful supplement treatments in the literature: [Dix77, Lax02].

The present chapter focused on the part of infinite-dimensional calculus often referred to as Malliavin calculus of variation.

E. Nelson [Nel69] studied representations of abelian algebras as multiplication operators in various L^2 spaces; equivalence, vs inequivalence, mutually singular measures and multiplicity theory. Representations of abelian algebras is often called multiplicity theory; or theory of spectral representations. The trick is that inequivalent representations corresponds to mutually singular measures. Chapter 6 of [op. cit] is one of the nicest accounts of multiplicity theory. This idea applies to our present analysis of representations of CCRs, because the representations of the CCRs restrict to representations of the dual pairs of abelian algebras.

6.1 Realization of the operators

> "*Quantum-attention functions are the keys to quantum machine learning.*" — Amit Ray, Quantum Computing Algorithms for Artificial Intelligence.

Below we make precise the aforementioned isometric isomorphism Z between the symmetric Fock space, based on a fixed Hilbert space $\mathscr{L}$, on the one hand, and the Gaussian path-space L^2 space on the other. Recall $\mathscr{L}$ allows a realization as a Gaussian Hilbert space, and the L^2 here refers to the associated probability space.

In Chapter 7 below, we shall further analyze key intertwining properties of the operator Z, see (6.1) below.

Theorem 6.1. *Let ω_{Fock} be the Fock state on $CCR(\mathscr{L})$, see (3.8)–(3.9), and let π_F denote the corresponding (Fock space) representation, acting on $\Gamma_{sym}(\mathscr{L})$, see Lemma 4.109. Let $Z : \Gamma_{sym}(\mathscr{L}) \longrightarrow L^2(\Omega, \mathbb{P})$ be the isomorphism given by*

$$Z\left(e^k\right) := e^{\Phi(k) - \frac{1}{2}\|k\|_{\mathscr{L}}^2}, \quad k \in \mathscr{L}. \tag{6.1}$$

Here $L^2(\Omega, \mathbb{P})$ denotes the Gaussian Hilbert space corresponding to $\mathscr{L}$; see Definition 4.103. For vectors $k \in \mathscr{L}$, let T_k denote the Malliavin derivative in the direction k; see Definition 5.2.

We then have the following realizations:

$$T_k = Z\pi_F\left(a(k)\right) Z^*, and \tag{6.2}$$

$$M_{\Phi(k)} - T_k = Z\pi_F\left(a^*(k)\right) Z^*; \tag{6.3}$$

valid for all $k \in \mathscr{L}$, where the two identities (6.2)–(6.3) hold on the dense domain $\mathscr{D}$ from Lemma 4.109.

Remark 6.2. The two formulas (6.2)–(6.3) take the following form, see Figs. 7.1–7.2.

In the proof of the theorem, we make use of the following:

Lemma 6.3. *Let* $\mathscr{L}$, $CCR(\mathscr{L})$, *and* ω_F (= *the Fock vacuum state*) *be as above. Then, for all* $n, m \in \mathbb{N}$, *and all* $h_1, \ldots, h_n, k_1, \ldots, k_m \in \mathscr{L}$, *we have the following identity:*

$$\omega_F\left(a\left(h_1\right)\ldots, a\left(h_n\right) a^*\left(k_m\right) \ldots a\left(k_1\right)\right)$$

$$= \delta_{n,m} \sum_{s \in S_n} \langle h_1, k_{s(1)} \rangle_{\mathscr{L}} \langle h_2, k_{s(2)} \rangle_{\mathscr{L}} \cdots \langle h_n, k_{s(n)} \rangle_{\mathscr{L}} \quad (6.4)$$

where the summation on the RHS in (6.4) *is over the symmetric group* S_n *of all permutations of* $\{1, 2, \ldots, n\}$. (*In the case of the CARs, the analogous expression on the RHS will instead be a determinant.*)

Proof. We leave the proof of the lemma to the reader; it is also contained in [BR81]. □

Remark 6.4. In physics-lingo, we say that the vacuum-state ω_F is determined by its *two-point functions*

$$\omega_F\left(a\left(h\right) a^*\left(k\right)\right) = \langle h, k \rangle_{\mathscr{L}}, \quad \text{and}$$

$$\omega_F\left(a^*\left(k\right) a\left(h\right)\right) = 0, \quad \forall h, k \in \mathscr{L}.$$

Proof of Theorem 6.1. We shall only give the details for formula (6.2). The modifications needed for (6.3) will be left to the reader.

6.2 The unitary group

For a given Gaussian field $(\mathscr{L}, \Omega, \mathcal{F}, \mathbb{P}, \Phi)$, we studied the CCR $(\mathscr{L})$-algebra, and the operators associated with its Fock-vacuum representation.

From the determination of Φ by

$$\mathbb{E}\left(e^{i\Phi(k)}\right) = e^{-\frac{1}{2}\|k\|_{\mathscr{L}}^2}, \quad k \in \mathscr{L}; \quad (6.5)$$

we deduce that $(\Omega, \mathcal{F}, \mathbb{P}, \Phi)$ satisfies the following covariance with respect to the group $\text{Uni}(\mathscr{L}) := G(\mathscr{L})$ of all unitary operators $U : \mathscr{L} \longrightarrow \mathscr{L}$.

We shall need the following:

Definition 6.5. We say that $\alpha \in Aut(\Omega, \mathcal{F}, \mathbb{P})$ if the following three conditions hold:

(i) $\alpha : \Omega \longrightarrow \Omega$ is defined $\mathbb{P}$ a.e. on Ω, and $\mathbb{P}(\alpha(\Omega)) = 1$.

(ii) $\mathcal{F} = \alpha(\mathcal{F})$; more precisely, $\mathcal{F} = \{\alpha^{-1}(B) \mid B \in \mathcal{F}\}$ where

$$\alpha^{-1}(B) = \{\omega \in \Omega \mid \alpha(\omega) \in B\}. \quad (6.6)$$

(iii) $\mathbb{P} = \mathbb{P} \circ \alpha^{-1}$, i.e., α is a measure preserving automorphism.

Note that when (i)–(iii) hold for α, then we have the unitary operators U_α in $L^2(\Omega, \mathcal{F}, \mathbb{P})$,

$$U_\alpha F = F \circ \alpha, \tag{6.7}$$

or more precisely,

$$(U_\alpha F)(\omega) = F(\alpha(\omega)), \quad \text{a.e. } \omega \in \Omega,$$

valid for all $F \in L^2(\Omega, \mathcal{F}, \mathbb{P})$.

Theorem 6.6.

(i) *For every $U \in G(\mathscr{L})$ ($=$the unitary group of $\mathscr{L}$), there is a unique $\alpha \in Aut(\Omega, \mathcal{F}, \mathbb{P})$ such that*

$$\Phi(Uk) = \Phi(k) \circ \alpha, \tag{6.8}$$

or equivalently (see (6.7))

$$\Phi(Uk) = U_\alpha(\Phi(k)), \quad \forall k \in \mathscr{L}. \tag{6.9}$$

(ii) *If $T : L^2(\Omega, \mathbb{P}) \longrightarrow L^2(\Omega, \mathbb{P}) \otimes \mathscr{L}$ is the Malliavin derivative from Definition 5.2, then we have:*

$$TU_\alpha = (U_\alpha \otimes U)T. \tag{6.10}$$

Proof. The first conclusion in the theorem is immediate from the above discussion, and we now turn to the covariance formula (6.10).

Note that (6.10) involves unbounded operators, and it holds on the dense subspace $\mathscr{D}$ in $L^2(\Omega, \mathbb{P})$ from Lemma 4.109. Hence it is enough to verify (6.10) on vectors in $L^2(\Omega, \mathbb{P})$ of the form $e^{\Phi(k) - \frac{1}{2}\|k\|^2_{\mathscr{L}}}$, $k \in \mathscr{L}$. Using Lemma 4.109, we then get:

$$\text{LHS}_{(6.10)}\left(e^{\Phi(k) - \frac{1}{2}\|k\|^2_{\mathscr{L}}}\right) = e^{-\frac{1}{2}\|k\|^2_{\mathscr{L}}} T\left(e^{\Phi(Uk)}\right) \text{(by (6.8))}$$

$$= e^{-\frac{1}{2}\|Uk\|^2_{\mathscr{L}}} e^{\Phi(Uk)} \otimes (Uk) \text{ (by Remark 5.3)}$$

$$= (U_\alpha \otimes U)\left(e^{\Phi(k) - \frac{1}{2}\|k\|^2_{\mathscr{L}}}\right)$$

$$= \text{RHS}_{(6.10)}$$

which is the desired conclusion. $\qquad\qquad\qquad\qquad\qquad\qquad\qquad\square$

6.3 The Fock-state, and representation of CCR, realized as Malliavin calculus

"The PCT refers to the combined symmetry of a quantum field theory under P Parity, C charge, and T time. "Spin and statistics" refers to the fact that, in quantum field theory, it can be proved that spin 1/2 particles obey Fermi–Dirac statistics, whereas integer spin 0, 1, 2 particles obey Bose–Einstein statistics." R. F. Streater and Arthur Wightman: *PCT, spin and statistics, and all that*, Princeton University Press, 1964.

We now resume our analysis of the representation of the canonical commutation relations (CCR)-algebra induced by the canonical Fock state (see (3.1)). In our analysis below, we shall make use of the following details: Brownian motion, It-integrals, and the Malliavin derivative.

By combining the present results in Chapters 5 and 6 we gain new insight into *Malliavin calculus*, and more generally into an extension, from finite to infinite dimensions, of the rules of calculus. Moreover, the link between CCR *representation theory*, on the one hand, and Malliavin calculus, on the other, is accomplished with the use of *intertwining operators*, studied in detail in Chapter 7 below; see especially Theorem 7.4. Indeed, the Malliavin derivative is a key notion in infinite-dimensional analysis. It is used in a variety of problems in stochastic calculus, in optimization, and in infinite-dimensional calculus of variation problems. One reason for the success of the Malliavin derivative is that it turns out to follow the familiar rules from calculus (most notable *integration by parts*), but now in the infinite-dimensional stochastic setting. For a more comprehensive study of Malliavin analysis, we refer to the books and papers cited below. Here we shall stress the following two interconnected features of the Malliavin derivative: (i) the correspondence to representations of the CCRs, and (ii) an infinite-dimensional integration-by part formula, and of Leibnitz' rule. For item (i), see especially Lemma 5.16 (Leibniz–Malliavin), Theorem 6.1, (6.18), Theorem 6.7, and Lemma 6.8, below. For (ii), see (6.22), and also Section 5.1.

The general setting. Let $\mathscr{L}$ be a fixed Hilbert space, and let $\mathrm{CCR}(\mathscr{L})$ be the $*$-algebra on the generators $a(k)$, $a^*(l)$, $k, l \in \mathscr{L}$, and subject to the relations for the CCR-algebra, see 3.1:

$$[a(k), a(l)] = 0, \quad \text{and} \tag{6.11}$$

$$[a(k), a^*(l)] = \langle k, l \rangle_{\mathscr{L}} \mathbb{1} \tag{6.12}$$

where $[\cdot, \cdot]$ is the commutator bracket.

A representation π of CCR $(\mathscr{L})$ consists of a fixed Hilbert space $\mathscr{H} = \mathscr{H}_\pi$ (the representation space), a dense subspace $\mathscr{D}_\pi \subset \mathscr{H}_\pi$, and a $*$-homomorphism $\pi : \text{CCR}\,(\mathscr{L}) \longrightarrow End\,(\mathscr{D}_\pi)$ such that

$$\mathscr{D}_\pi \subset dom(\pi(A)), \quad \forall A \in \text{CCR}. \tag{6.13}$$

The representation axiom entails the commutator properties resulting from (6.11)–(6.12); in particular π satisfies

$$[\pi(a(k)), \pi(a(l))]F = 0, \quad \text{and} \tag{6.14}$$

$$[\pi(a(k)), \pi(a(l))^*]F = \langle k, l \rangle_{\mathscr{L}} F, \tag{6.15}$$

$\forall k, l \in \mathscr{L}$, $\forall F \in \mathscr{D}_\pi$; where $\pi\,(a^*\,(l)) = \pi\,(a\,(l))^*$.

In the application below, we take $\mathscr{L} = L^2\,(0, \infty)$, and $\mathscr{H}_\pi = L^2\,(\Omega, \mathcal{F}_\Omega, \mathbb{P})$ where $(\Omega, \mathcal{F}_\Omega, \mathbb{P})$ is the standard Wiener probability space, and

$$\Phi_t(\omega) = \omega(t), \quad \forall \omega \in \Omega, \; t \in [0, \infty). \tag{6.16}$$

For $k \in \mathscr{L}$, we set

$$\Phi\,(k) = \int_0^\infty k\,(t)\,d\Phi_t \quad (= \text{the Itô-integral.})$$

The dense subspace $\mathscr{D}_\pi \subset \mathscr{H}_\pi$ is generated by the polynomial fields:

For $n \in \mathbb{N}$, $h_1, \ldots, h_n \in \mathscr{L} = L^2_{\mathbb{R}}\,(0, \infty)$, $p \in \mathbb{R}^n \longrightarrow \mathbb{R}$ a polynomial in n real variables, set

$$F = p(\Phi(h_1), \ldots, \Phi(h_n)), \quad \text{and} \tag{6.17}$$

$$\pi\,(a\,(k))\,F = \sum_{j=1}^n \left(\frac{\partial}{\partial x_j}p\right)(\Phi(h_1), \ldots, \Phi(h_n))\,\langle h_j, k \rangle. \tag{6.18}$$

It follows from Lemma 5.16 that $\mathscr{D}_\pi$ is an algebra under pointwise product and that

$$\pi\,(a\,(k))\,(FG) = (\pi\,(a\,(k))\,F)\,G + F\,(\pi\,(a\,(k))\,G), \tag{6.19}$$

$\forall k \in \mathscr{L}$, $\forall F, G \in \mathscr{D}_\pi$. Equivalently, $T_k := \pi\,(a\,(k))$ is a derivation in the algebra $\mathscr{D}_\pi$ (relative to pointwise product.)

Theorem 6.7. *With the operators* $\pi\,(a\,(k))$, $k \in \mathscr{L}$, *we get a* $*$-*representation* $\pi : \text{CCR}\,(\mathscr{L}) \longrightarrow End\,(\mathscr{D}_\pi)$, *i.e.,* $\pi\,(a\,(k)) = $ *the Malliavin derivative in the direction* k,

$$\pi\,(a\,(k))\,F = \langle T(F), k \rangle_{\mathscr{L}}, \quad \forall F \in \mathscr{D}_\pi, \forall k \in \mathscr{L}. \tag{6.20}$$

Proof. The proof begins with the following lemma. □

Lemma 6.8. *Let π, $CCR(\mathscr{L})$, and $\mathscr{H}_\pi = L^2(\Omega, \mathcal{F}_\Omega, \mathbb{P})$ be as above. For $k \in \mathscr{L}$, we shall identify $\Phi(k)$ with the unbounded multiplication operator in $\mathscr{H}_\pi$:*

$$\mathscr{D}_\pi \ni F \longmapsto \Phi(k) F \in \mathscr{H}_\pi. \tag{6.21}$$

For $F \in \mathscr{D}_\pi$, we have $\pi(a(k))^* F = -\pi(a(k)) F + \Phi(k) F$; or in abbreviated form:

$$\pi(a(k))^* = -\pi(a(k)) + \Phi(k) \tag{6.22}$$

valid on the dense domain $\mathscr{D}_\pi \subset \mathscr{H}_\pi$.

Proof. This follows from the following computation for $F, G \in \mathscr{D}_\pi$, $k \in \mathscr{L}$. Setting $T_k := \pi(a(k))$, we have

$$\mathbb{E}(T_k(F) G) + \mathbb{E}(F T_k(G)) = \mathbb{E}(T_k(FG)) = \mathbb{E}(\Phi(k)FG).$$

Hence $\mathscr{D}_\pi \subset dom(T_k^*)$, and $T_k^*(F) = -T_k(F) + \Phi(k) F$, which is the desired conclusion (6.22). □

Proof of Theorem 6.7 continued. It is clear that the operators $T_k = \pi(a(k))$ form a commuting family. Hence on $\mathscr{D}_\pi$, we have for $k, l \in \mathscr{L}$, $F \in \mathscr{D}_\pi$:

$$
\begin{aligned}
[T_k, T_l^*](F) &= [T_k, \Phi(l)](F) && \text{by (6.22)}\\
&= T_k(\Phi(l) F) - \Phi(l)(T_k(F)) \\
&= T_k(\Phi(l)) F && \text{by (6.19)}\\
&= \langle k, l \rangle_{\mathscr{L}} F && \text{by (6.18)}
\end{aligned}
$$

which is the desired commutation relation (6.12).

The remaining check on the statements in the theorem are now immediate. □

Corollary 6.9. *The state on $CCR(\mathscr{L})$ which is induced by π and the constant function $\mathbb{1}$ in $L^2(\Omega, \mathbb{P})$ is the Fock-vacuum-state, ω_{Fock}.*

Proof. The assertion will follow once we verify the following two conditions:

$$\int_\Omega T_k^* T_k(\mathbb{1}) \, d\mathbb{P} = 0 \tag{6.23}$$

and

$$\int_\Omega T_k T_l^* \left(\mathbb{1}\right) d\mathbb{P} = \langle k, l \rangle_{\mathscr{L}} \qquad (6.24)$$

for all $k, l \in \mathscr{L}$.

This in turn is a consequence of our discussion of (3.8)–(3.9) above: The Fock state ω_{Fock} is determined by these two conditions. The assertions (6.23)–(6.24) follow from $T_k \left(\mathbb{1}\right) = 0$, and $\left(T_k T_l^*\right)\left(\mathbb{1}\right) = \langle k, l \rangle_{\mathscr{L}} \mathbb{1}$. See (5.13). $\qquad \square$

Corollary 6.10. *For $k \in L^2_{\mathbb{R}}(0, \infty)$ we get a family of selfadjoint multiplication operators $T_k + T_k^* = M_{\Phi(k)}$ on $\mathscr{D}_\pi$ where $T_k = \pi\left(a\left(k\right)\right)$. Moreover, the von Neumann algebra generated by these operators is $L^\infty\left(\Omega, \mathbb{P}\right)$, i.e., the maximal abelian L^∞-algebra of all multiplication operators in $\mathscr{H}_\pi = L^2\left(\Omega, \mathbb{P}\right)$.*

Remark 6.11. In our considerations of representations π of CCR$\left(\mathscr{L}\right)$ in a Hilbert space $\mathscr{H}_\pi$, we require the following five axioms satisfied:

(i) a dense subspace $\mathscr{D}_\pi \subset \mathscr{H}_\pi$;
(ii) $\pi : \text{CCR}\left(\mathscr{L}\right) \longrightarrow End\left(\mathscr{D}_\pi\right)$, i.e., $\mathscr{D}_\pi \subset \cap_{A \in \text{CCR}(\mathscr{L})} dom\left(\pi\left(A\right)\right)$;
(iii) $[\pi(a(k)), \pi\left(a\left(l\right)\right)] = 0, \forall k, l \in \mathscr{L}$;
(iv) $\left[\pi(a(k)), \pi\left(a\left(l\right)\right)^*\right] = \langle k, l \rangle_{\mathscr{L}} I_{\mathscr{H}_\pi}, \forall k, l \in \mathscr{L}$; and
(v) $\pi\left(a^*\left(k\right)\right) \subset \pi\left(a\left(k\right)\right)^*, \forall k \in \mathscr{L}$.

Note that in our assignment for the operators $\pi\left(a\left(k\right)\right)$, and $\pi\left(a^*\left(k\right)\right)$ in Lemma 6.8, we have all the conditions (i)–(iv) satisfied. We say that π is a *selfadjoint representation*.

If alternatively, we define

$$\rho : \text{CCR}\left(\mathscr{L}\right) \longrightarrow End\left(\mathscr{D}_\pi\right) \qquad (6.25)$$

with the following modification:

$$\begin{cases} \rho\left(a\left(k\right)\right) = T_k, \ k \in \mathscr{L}, \quad \text{and} \\ \rho\left(a^*\left(k\right)\right) = \Phi\left(k\right) \end{cases} \qquad (6.26)$$

then this ρ will satisfy (i)–(iii), and

$$[\rho(a(k)), \rho\left(a^*\left(l\right)\right)] = \langle k, l \rangle_{\mathscr{L}} I_{\mathscr{H}_\pi};$$

but then $\rho\left(a\left(k\right)\right) \subsetneqq \rho\left(a\left(k\right)\right)^*$; i.e., non-containment of the respective graphs.

One generally says that the representation π is (formally) selfadjoint, while the second representation ρ is *not*.

6.4 Conclusions: The general case

Below we give a brief overview of the correspondence between *Malliavin calculus*, on the one hand, and the study of *representations of the CCRs*, on the other. In Chapter 7 we outline how this can be made precise with the use of *intertwining operators*.

Definition 6.12. A representation π of CCR $(\mathscr{L})$ is said to be *admissible* if $\exists\,(\Omega, \mathcal{F}, \mathbb{P})$ as above such that $\mathscr{H}_\pi = L^2\,(\Omega, \mathcal{F}, \mathbb{P})$, and there exists a linear mapping $\Phi : \mathscr{L} \longrightarrow L^2\,(\Omega, \mathcal{F}, \mathbb{P})$ subject to the condition:

For every $n \in \mathbb{N}$, and every $k, h_1, \ldots, h_n \in \mathscr{L}$, the following holds on its natural dense domain in $\mathscr{H}_\pi$: For every $p \in \mathbb{R}\,[x_1, \ldots, x_n]$, we have

$$
\pi\left(\left[a(k), p\left(a^*(h_1), \ldots, a^*\left(h_n\right)\right)\right]\right) = \sum_{i=1}^{n} \langle k, h_i \rangle_{\mathscr{L}}\, M_{\frac{\partial p}{\partial x_i}\left(\Phi(h_1), \ldots, \Phi(h_n)\right)},
$$

$$(6.27)$$

with the M on the RHS denoting "multiplication."

Corollary 6.13.

(i) *Every admissible representation π of CCR $(\mathscr{L})$ yields an associated Malliavin derivative as in (6.27).*

(ii) *The Fock-vacuum representation π_F is admissible.*

Proof. (6.13) follows from the definition combined with Corollary 2.7. (6.13) is a direct consequence of Lemma 5.7 and Theorem 5.8; see also Corollary 6.9. $\qquad\square$

6.5 Guide to the literature

While we have cited some paper from the Bibliography in our discussion inside the chapter, readers not familiar with the ideas involved may wish to consult the following sources: [BR81, GJ87, HKPS13, LG14, LS08].

Chapter 7

Intertwining Operators and Their Realizations in Stochastic Analysis

> *"Representation theory is pervasive across fields of mathematics for two reasons. First, the applications of representation theory are diverse: in addition to its impact on algebra, representation theory, and it illuminates and generalizes Fourier analysis via harmonic analysis"* T. Y. Lam, (1998), "Representations of finite groups: a hundred years", Notices of the AMS, 45.

The study of intertwining operators is a key ingredient of representation theory, and it has diverse settings and applications, and many and diverse applications. For our present purposes we have made choices as follows, each motivated by the aim of the book: Representations derived from states via the GNS construction.

7.1 Representations from states, the GNS construction, and intertwining operators

The link between Gaussian processes (from duality theory) on the one hand (Chapter 4), and representations of the CCRs (Chapter 3), on the other allows us to give a display new families of Gaussian processes, and to find explicit formulas and transforms. This is outlined below with the use of representation theory and duality tools in such a way that the corresponding Gaussian path-space measures are mutually singular. The link to representation theory makes use of a corresponding family of representations of the *canonical commutation relations* (CCR) in an infinite number of degrees of freedom.

A key feature of our construction is the use of *intertwining operators* and explicit formulas for associated transforms; the latter are *infinite-dimensional analogues of Fourier transforms*. Our framework is that of Gaussian Hilbert spaces, reproducing kernel Hilbert spaces, and the corresponding Fock spaces. This then forms the setting for our CCR representations. We further show, with the use of representation theory and infinite-dimensional analysis, that our pairwise inequivalent probability spaces (for the Gaussian processes) correspond in an explicit manner to pairwise disjoint CCR representations.

In the previous chapter we introduced a class of representations (by operators in Hilbert space.) In general, there are intriguing varieties of representations available for a given problem at hand. How do we decide whether or not two representations are equivalent, or whether a sub-representation of one is equivalent to a sub-representation of the other. For this purpose we must study what is called intertwining operators, they will be defined and analyzed below.

Our discussion in Chapters 4 through 6 above, entailed a variety of representations. One is naturally led to ask for criteria for making distinctions between a pair or more representations. For this purpose we introduce and study, below, the notion of *intertwining operators*; see, e.g., Equation (7.5).

Definition 7.1. Let $\mathfrak{A}$ be a $*$-algebra, and let ω, ψ be two states. We say that $\omega \leq \psi$ if and only if

$$\omega\left(A^*A\right) \leq \psi\left(A^*A\right), \quad \forall A \in \mathfrak{A}. \tag{7.1}$$

Lemma 7.2. *Given two states* ω, ψ *on* $\mathfrak{A}$, *let* $(\pi_\omega, v_\omega, \mathscr{H}(\omega))$ *and* $(\pi_\psi, v_\psi, \mathscr{H}(\psi))$ *be the respective GNS representations. Then we have a bounded intertwining operator* $L : \mathscr{H}(\psi) \to \mathscr{H}(\omega)$ *if and only if* $\omega \leq \psi$.

Proof. Assume we have two states ω, ψ and $L : \mathscr{H}(\psi) \to \mathscr{H}(\omega)$ satisfying

$$L\pi_\psi\left(A\right) = \pi_\omega\left(A\right) L \tag{7.2}$$

for all $A \in \mathfrak{A}$, i.e., L is intertwining for the two representations. From the definition of the GNS-construction, for all $A \in \mathfrak{A}$, we have

$$\omega\left(A\right) = \langle v_\omega, \pi_\omega(A)v_\omega \rangle_{\mathscr{H}(\omega)},$$
$$\psi\left(A\right) = \langle v_\psi, \pi_\psi(A)v_\psi \rangle_{\mathscr{H}(\psi)},$$

and so

$$L\left(v_\psi\right) = v_\omega \tag{7.3}$$

where L is the intertwining operator from (7.2).

We now prove that (7.1) holds: Note that the space

$$\{\pi_\omega\left(A\right) v_\omega : A \in \mathfrak{A}\} \tag{7.4}$$

is dense in $\mathscr{H}\left(\omega\right)$ from the definition of the GNS-representation.

Since $L : \mathscr{H}\left(\psi\right) \to \mathscr{H}\left(\omega\right)$ is assumed bounded, it is enough to evaluate it on the vectors in (7.4). we now turn on (7.1). Let $A \in \mathfrak{A}$, then

$$
\begin{aligned}
\omega\left(A^*A\right) &= \langle v_\omega, \pi_\omega\left(A^*A\right) v_\omega \rangle_{\mathscr{H}(\omega)} \\
&= \langle \pi_\omega\left(A\right) v_\omega, \pi_\omega\left(A\right) v_\omega \rangle_{\mathscr{H}(\omega)} \\
&= \|\pi_\omega\left(A\right) v_\omega\|^2_{\mathscr{H}(\omega)} \\
&\overset{\text{by } (7.3)}{=} \|L\pi_\psi\left(A\right) v_\psi\|^2_{\mathscr{H}(\psi)} \\
&\leq \|L\|^2 \|\pi_\psi(A) v_\psi\|^2_{\mathscr{H}(\psi)},
\end{aligned}
$$

where $\|L\|$ denotes the operator norm, and L is viewed as a linear operator $L : \mathscr{H}\left(\psi\right) \to \mathscr{H}\left(\omega\right)$, so $\|L\| = \|L\|_{\mathscr{H}(\psi) \to \mathscr{H}(\omega)} \leq 1$, since L was assumed contractive. Hence $\omega\left(A^*A\right) \leq \psi\left(A^*A\right) = \|\pi_\psi\left(A\right) v_\psi\|^2_{\mathscr{H}(\psi)}$ holds, which is the desired Inequality (7.1).

The reader will notice that the above argument works in reversal, and so the lemma is proved. $\qquad\square$

Corollary 7.3. *A GNS representation π_ψ (for a state ψ on $\mathfrak{A}$) is irreducible if and only if the only states ω on $\mathfrak{A}$ satisfying (7.1), i.e., $\omega\left(A^*A\right) \leq \psi\left(A^*A\right)$, $\forall A \in \mathfrak{A}$, are $\omega = \psi$. In this case we say that ψ is a pure state. Hence we get the following: The representation ψ_π is irreducible if and only if the state ψ is pure.*

Theorem 7.4. *The operator Z in (6.1) is an isomorphic isomorphism, i.e., a unitary operator from $\Gamma_{sym}\left(\mathscr{L}\right)$ onto $L^2\left(\Omega, \mathbb{P}\right)$, such that*

$$T_k Z = Z \pi_F\left(a\left(k\right)\right) \tag{7.5}$$

holds on the dense subspace of all finite symmetric tensor polynomials in $\Gamma_{sym}\left(\mathscr{L}\right)$; or equivalently on the dense subspace in $\Gamma_{sym}\left(\mathscr{L}\right)$ spanned by

$$\Gamma\left(l\right) := e^l := \sum_{n=0}^{\infty} \frac{l^{\otimes n}}{\sqrt{n!}} \in \Gamma_{sym}\left(\mathscr{L}\right), \, l \in \mathscr{L}; \tag{7.6}$$

see Figures 7.1–7.2, also Lemma 4.109.

$$
\begin{CD}
\Gamma_{sym}(\mathscr{L}) @>Z>> L^2(\Omega,\mathbb{P}) \\
@V\pi_F(a(k))VV @VVT_kV \\
\Gamma_{sym}(\mathscr{L}) @>>Z> L^2(\Omega,\mathbb{P})
\end{CD}
$$

Figure 7.1: The first operator.

$$
\begin{CD}
\Gamma_{sym}(\mathscr{L}) @>Z>> L^2(\Omega,\mathbb{P}) \\
@V\pi_F(a^*(k))VV @VVM_{\Phi(k)}-T_kV \\
\Gamma_{sym}(\mathscr{L}) @>>Z> L^2(\Omega,\mathbb{P})
\end{CD}
$$

Figure 7.2: The second operator.

Proof. We now compute (7.5) on the vectors e^l in (7.6):

$$
\begin{aligned}
T_k Z\left(e^l\right) &= T_k\left(e^{\Phi(l)-\frac{1}{2}\|l\|^2_{\mathscr{L}}}\right) && \text{(by Lemma 4.109)}\\
&= e^{-\frac{1}{2}\|l\|^2_{\mathscr{L}}} T_k\left(e^{\Phi(l)}\right) \\
&= e^{-\frac{1}{2}\|l\|^2_{\mathscr{L}}} \langle k,l\rangle_{\mathscr{L}}\, e^{\Phi(l)} && \text{(by Remark 5.3)}\\
&= Z\pi_F(a(k))(e^l),
\end{aligned}
$$

valid for all $k,l \in \mathscr{L}$. $\qquad\square$

Remark 7.5. Figure 7.2 offers a picture version of the visual meaning of the notion of "intertwining" for a pair of representations. The reader is encouraged to fill in the details in verifying the property for Z reflected in Figure 7.2. For additional details, see also Theorem 6.1 above, and its proof.

7.2 Guide to the literature

While we have cited some paper from the Bibliography in our discussion inside the chapter, readers not familiar with the ideas involved may wish to consult the following sources: [GJ87, Hid80, Jan97].

In this paper here [Sti55] by Stinespring, the notion, and the key result, for completely positive maps on C^*-algebras are introduced. The subject was subsequently taken up, and greatly expanded, by numerous authors, and pioneered by W. Arveson.

Chapter 8

Applications

> *"Pure mathematics is much more than an armoury of tools and techniques for the applied mathematician. On the other hand, the pure mathematician has ever been grateful to applied mathematics for stimulus and inspiration. From the vibrations of the violin string they have drawn enchanting harmonies of Fourier Series, and to study the triode valve they have invented a whole theory of non-linear oscillations."* George Frederick James Temple, 100 Years of Mathematics: a Personal Viewpoint (1981).

While the diversity of applications is vast, in this concluding chapter we have selected the following: Machine learning and choices of feature spaces, optimal sampling tools, infinite network models and their potential theory.

While we studied Gaussian processes, Gaussian fields, Itô calculus, Malliavin derivatives, stochastic differential equations, and stochastic PDEs in earlier chapters, the present chapter will discuss (a few) applications.

The last section deals with the following general question for p.d. kernels K. In a general context of sampling, one often seeks to decide whether or not the Dirac point functions δ_x are in $\mathscr{H}(K)$, the RKHS. This occurs, for example in the following context, arising in many applied questions amenable to kernel analysis. Consider, for example, the framework of continuous domains, e.g., the case of p.d. kernels K defined on $S \times S$ where S is a domain in some $\mathbb{R}^n$. For these questions, it is important to devise optimal sampling strategies. A choice of sampling points then leads to useful models where the base set S will be discrete. In order to adapt various optimization strategies, it is therefore important for one to be able to easily decide whether or not the Dirac point functions δ_x are in $\mathscr{H}(K)$, (the RKHS) or not. We give an answer to the question in Theorem 8.14 below.

8.1 Machine learning

One of the more recent applications of kernels and the associated reproducing kernel Hilbert spaces (RKHS) is to optimization, also called kernel-optimization. See, e.g., [YLTL18, LLL11, BSS90, BSS89]. In the context of machine learning, it refers to training-data and feature spaces. In the context of numerical analysis, a popular version of the method is used to produce splines from sample points; and to create best spline-fits. In statistics, there are analogous optimization problems going by the names "least-square fitting," and "maximum-likelihood" estimation. In the latter instance, the object to be determined is a suitable probability distribution which makes "most likely" the occurrence of some data which arises from experiments, or from testing.

What these methods have in common is a minimization (or a max problem) involving a "quadratic" expression Q with two terms. The first in Q measures a suitable $L^2(\mu)$-square applied to a difference of a measurement and a "best fit." The latter will then to be chosen from anyone of a number of suitable reproducing kernel Hilbert spaces (RKHS). The choice of kernel and RKHS will serve to select desirable features. So we will minimize a quantity Q which is the sum of two terms as follows: (i) a L^2-square applied to a difference, and (ii) a penalty term which is a RKHS norm-squared. (See (8.2).) In the application to determination of splines, the penalty term may be a suitable Sobolev normed-square; i.e., L^2 norm-squared applied to a chosen number of derivatives. Hence non-differentiable choices will be "penalized."

In all of the cases, discussed above, there will be a good choice of (i) and (ii), and we show that there is then an explicit formula for the optimal solution; see (8.5) in Theorem 8.1 below.

Let X be a set, and let $K : X \times X \longrightarrow \mathbb{C}$ be a positive definite (p.d.) kernel. Let $\mathcal{H}(K)$ be the corresponding reproducing kernel Hilbert space (RKHS). Let $\mathscr{B}$ be a sigma-algebra of subsets of X, and let μ be a positive measure on the corresponding measure space $(X, \mathscr{B})$. We assume that μ is sigma-finite. We shall further assume that the associated operator T given by

$$\mathcal{H}(K) \ni f \xrightarrow{\ T\ } (f(x))_{x \in X} \in L^2(\mu) \tag{8.1}$$

is densely defined and closable.

Fix $\beta > 0$, and $\psi \in L^2(\mu)$, and set

$$Q_{\psi,\beta}(f) = \|\psi - Tf\|^2_{L^2(\mu)} + \beta \|f\|^2_{\mathcal{H}(K)} \qquad (8.2)$$

defined for $f \in \mathcal{H}(K)$, or in the dense subspace $dom(T)$ where T is the operator in (8.1). Let

$$L^2(\mu) \xrightarrow{T^*} \mathcal{H}(K) \qquad (8.3)$$

be the corresponding adjoint operator, i.e.,

$$\langle F, T^*\psi \rangle_{\mathcal{H}(K)} = \langle Tf, \psi \rangle_{L^2(\mu)} = \int_X \overline{f(s)}\psi(s)\, d\mu(s). \qquad (8.4)$$

Theorem 8.1. *Let K, μ, ψ, β be as specified above; then the optimization problem*

$$\inf_{f \in \mathcal{H}(K)} Q_{\psi,\beta}(f)$$

has a unique solution F in $\mathcal{H}(K)$, it is

$$F = (\beta I + T^*T)^{-1} T^*\psi \qquad (8.5)$$

where the operator T and T^* are as specified in (8.1)–(8.4).

Proof. (Sketch) We fix F, and assign $f_\varepsilon := F + \varepsilon h$ where h varies in the dense domain $dom(T)$ from (8.1). For the derivative $\frac{d}{d\varepsilon}\big|_{\varepsilon=0}$ we then have:

$$\frac{d}{d\varepsilon}\Big|_{\varepsilon=0} Q_{\psi,\beta}(f_\varepsilon) = 2\Re \langle h, (\beta I + T^*T) F - T^*\psi \rangle_{\mathcal{H}(K)} = 0$$

for all h in a dense subspace in $\mathcal{H}(K)$. The desired conclusion follows. $\square$

Least-square Optimization

We now specialize the optimization formula from Theorem 8.1 to the problem of minimize a "quadratic" quantity Q. It is still the sum of two individual terms: (i) a L^2-square applied to a difference, and (ii) a penalty term which is the RKHS norm-squared. But the least-square term in (i) will simply be a sum of a finite number of squares of differences; hence "least-squares." As an application, we then get an easy formula (Theorem 8.2) for the optimal solution.

Let K be a positive definite kernel on $X \times X$ where X is an arbitrary set, and let $\mathcal{H}(K)$ be the corresponding reproducing kernel Hilbert space

(RKHS). Let $m \in \mathbb{N}$, and consider sample points:

$\{t_j\}_{j=1}^m$ as a finite subset in X, and

$\{y_i\}_{i=1}^m$ as a finite subset in $\mathbb{R}$, or equivalently, a point in $\mathbb{R}^m$.

Fix $\beta > 0$, and consider $Q = Q_{(\beta, t, y)}$, defined by

$$Q(f) = \underbrace{\sum_{i=1}^m |f(t_i) - y_i|^2}_{\text{least square}} + \underbrace{\beta \, \|f\|_{\mathcal{H}(K)}^2}_{\text{penalty form}}, \quad f \in \mathcal{H}(K). \qquad (8.6)$$

We introduce the associated dual pair of operators as follows:

$$T : \mathcal{H}(K) \longrightarrow \mathbb{R}^m \simeq l_m^2, \quad \text{and}$$
$$T^* : l_m^2 \longrightarrow \mathcal{H}(K) \qquad (8.7)$$

where

$$Tf = (f(t_i))_{i=1}^m, \quad f \in \mathcal{H}(K); \quad \text{and} \qquad (8.8)$$

$$T^* y = \sum_{i=1}^m y_i K(\cdot, t_i) \in \mathcal{H}(K), \qquad (8.9)$$

for all $\vec{y} = (y_i) \in \mathbb{R}^m$.

Note that the duality then takes the following form:

$$\langle T^* y, f \rangle_{\mathcal{H}(K)} = \langle y, Tf \rangle_{l_m^2}, \quad \forall f \in \mathcal{H}(K), \quad \forall y \in l_m^2; \qquad (8.10)$$

consistent with (8.4).

Applying Theorem 8.1 to the counting measure

$$\mu = \sum_{i=1}^m \delta_{t_i} = \delta_{\{t_i\}}$$

for the set of sample points $\{t_i\}_{i=1}^m$, we get the two formulas:

$$T^* T f = \sum_{i=1}^m f(t_i) K(\cdot, t_i) = \sum_{i=1}^m f(t_i) K_{t_i}, \quad \text{and} \qquad (8.11)$$

$$TT^* y = K_m \vec{y} \qquad (8.12)$$

where K_m denotes the $m \times m$ matrix

$$K_m = (K(t_i, t_j))_{i,j=1}^m = \begin{pmatrix} K(t_1, t_1) & \cdots & \cdots & K(t_1, t_m) \\ K(t_2, t_1) & \cdots & \cdots & K(t_2, t_m) \\ \vdots & & & \\ K(t_m, t_1) & & & K(t_m, t_m) \end{pmatrix}. \qquad (8.13)$$

Theorem 8.2. *Let K, X, $\{t_i\}_{i=1}^m$, and $\{y_i\}_{i=1}^m$ be as above, and let K_m be the induced sample matrix (8.13).*

Fix $\beta > 0$; consider the optimization problem with

$$Q_{\beta,\{t_i\},\{y_i\}}(f) = \sum_{i=1}^m |y_i - f(t_i)|^2 + \beta\|f\|^2_{\mathcal{H}(K)}, \quad f \in \mathcal{H}(K). \quad (8.14)$$

Then the unique solution to (8.14) is given by

$$F(\cdot) = \sum_{i=1}^m (K_m + \beta I_m)_i^{-1} K(\cdot, t_i) \quad on\ X; \quad (8.15)$$

i.e., $F = \arg\min Q$ on $\mathcal{H}(K)$.

Proof. From Theorem 8.1, we get that the unique solution $F \in \mathcal{H}(K)$ is given by:

$$\beta F + T^* T F = T^* y,$$

and by (8.11)–(8.12), we further get

$$\beta F(\cdot) = \sum_{i=1}^m (y_i - F(t_i)) K(\cdot, t_i) \quad (8.16)$$

where the dot $\cdot$ refers to a free variable in X. An evaluation of (8.16) on the sample points yields:

$$\beta \vec{F} = K_m \left(\vec{y} - \vec{F} \right) \quad (8.17)$$

where $\vec{F} := (F(t_i))_{i=1}^m$, and $\vec{y} = (y_i)_{i=1}^m$. Hence

$$\vec{F} = (\beta I_m + K_m)^{-1} K_m \vec{y}. \quad (8.18)$$

Now substitute (8.18) into (8.17), and the desired conclusion in the theorem follows. We used the matrix identity

$$I_m - (\beta I_m + K_m)^{-1} K_m = \beta (\beta I_m + K_m)^{-1}. \qquad \square$$

8.2 Relative RKHSs

In previous chapters, we discussed two cases, positive definite (p.d.) kernels K, and the analogous conditionally negative definite (CND) kernels N. The first case is covered in detail, and it is usually connected with the name Aronszajn [Aro50, AS57, AAH69], while the second with Schoenberg [Sch37, Sch38, vNS41, Sch42].

The setting in both cases is a set S and the Cartesian product $S \times S$, and the two kernel functions K and N are defined on $S \times S$. Results for the p.d. case usually extend relatively easily, arguments modify so we get analogous results for the CND. The CND case is more general and cover many network models. There is also a close connection between the two cases, and most of the results there are usually credited to Schoenberg. From the comments below it is clear that the CND theory includes the more familiar p.d. theory, the latter usually named after Aronszajn.

In both cases, we define a vector space, so different spaces for the two kernels K and N. They become pre-Hilbert spaces with inner product definitions involving K and N. For N we must condition coefficients to have sum 0.

Let's denote the respective Hilbert completions $\mathscr{H}(K)$ and $\mathscr{H}(N)$. For $f \in \mathscr{H}(K)$ the values $f(s)$ are reproduced with $K(s, \cdot)$. For $g \in \mathscr{H}(N)$ the values of differences $g(s) - g(t)$ are reproduced with $N(s, \cdot) - N(t, \cdot)$. Note that $N(s, \cdot) - N(t, \cdot)$ is in $\mathscr{H}(N)$ but $N(s, \cdot)$ is not. So the two cases are different.

The network models are of the CND kind, and differences are reproduced, but not values. This is consistent with voltage differences. One cannot measure voltage itself, only differences. Of course one can ground the electric network (set ground equal to $V = 0$), and measure the voltage difference between an instrument and the ground.

In the first named author's earlier work on electrical networks and CND kernels (and graph Laplacians) the kernel which represents voltage differences for vertices s and t is called a *dipole*. We typically do not have monopoles, only dipoles. So the dipoles $V_{s,t}$ are in $\mathscr{H}(N)$. In our work, $\mathscr{H}(N)$ is realized as a finite energy Hilbert space. For more details, see, e.g., [JP08, JP14, JP13, JP11b, JP11a, JP11c, JP10], and [JT18a].

In this section, we study infinite networks with the use of frames in Hilbert space. While these results apply to finite systems, we will concentrate on the infinite case because of its statistical significance. By a network we mean a graph G with vertices V and edges E. We assume that each vertex is connected by single edges to a finite number of neighboring vertices, and that resistors are assigned to the edges. From this we define an associated graph-Laplacian Δ, and a resistance metric on V.

The functions on V of interest represent voltage distributions. While there are a number of candidates for associated Hilbert spaces of functions on V, the one we choose here has its norm-square equal to the energy of the voltage function. This Hilbert space is denoted $\mathscr{H}_E$, and it depends on an assigned conductance function (i.e., reciprocal of resistance.)

We may identify a canonical Parseval frame in $\mathscr{H}_E$, and we show that it is not an orthonormal basis except in simple degenerate cases. The frame vectors for $\mathscr{H}_E$ are indexed by oriented edges e, a dipole vector for each e, and a current through e.

The starting point is a fixed conductance function c for G. It is known that, to every prescribed conductance function c on the edges E of G, there is an associated pair $(\mathscr{H}_E, \Delta)$ where $\mathscr{H}_E$ in an energy Hilbert space, and $\Delta \, (= \Delta_c)$ is the c-Graph Laplacian; both depending on the choice of conductance function c.

We begin with the basic notions. Let V be a countable discrete set, and let $E \subset V \times V$ be a subset such that:

(i) $(x, y) \in E \iff (y, x) \in E$; $x, y \in V$;
(ii) $\#\{y \in V \mid (x, y) \in E\}$ is finite, and > 0 for all $x \in V$;
(iii) $(x, x) \notin E$; and
(iv) the networks is assumed connected, in the sense that, $\exists o \in V$ such that for all $y \in V$ $\exists x_0, x_1, \ldots, x_n \in V$ with $x_0 = o$, $x_n = y$, $(x_{i-1}, x_i) \in E$, $\forall i = 1, \ldots, n$.

Definition 8.3. A function $c : E \to \mathbb{R}_+ \cup \{0\}$ is called *conductance function*, if $c(e) \geq 0$, $\forall e \in E$, and if for all $x \in V$, and $(x, y) \in E$, $c_{xy} > 0$, and

$$c_{xy} = c_{yx}.$$

Given $x \in V$, let

$$c(x) := \sum_{y \sim x} c_{xy}, \tag{8.19}$$

where the summation is over y such that $(x, y) \in E$, and this is denoted by $y \sim x$.

Definition 8.4. When c is a conductance function, the corresponding Graph Laplacian $\Delta \, (= \Delta_c)$ is specified as

$$(\Delta u)(x) = \sum_{y \sim x} c_{xy} \, (u(x) - u(y)) \tag{8.20}$$

$$= c(x) u(x) - \sum_{y \sim x} c_{xy} u(y).$$

Reversible Markov process If (V, E, c) is given as in Definition 8.4, then for $(x, y) \in E$, set

$$p_{xy} := \frac{c_{xy}}{c(x)} \tag{8.21}$$

and note then $\{p_{xy}\}$ in (8.21) is a system of transition probabilities, i.e., $\sum_y p_{xy} = 1$, $\forall x \in V$, see Figure 8.1 below.

Definition 8.5. A Markov-random walk on V with transition probabilities (p_{xy}) is said to be *reversible* if $\exists$ a positive function $\widetilde{c}$ on V such that

$$\widetilde{c}(x) \, p_{xy} = \widetilde{c}(y) \, p_{yx}, \tag{8.22}$$

for all $(xy) \in E$.

Lemma 8.6. *There is a bijective correspondence between reversible Markov-walks on the one hand, and conductance functions on the other.*

Proof. If c is a conductance function on E, then (p_{xy}) defined in (8.21) is a reversible walk. This follows from $c_{xy} = c_{yx}$.

Conversely, if (8.22) holds for a system of transition probabilities

$$p_{xy} = \text{Prob}(x \mapsto y),$$

then $c_{xy} := \widetilde{c}(x) \, p_{xy}$ is a conductance function, where

$$\widetilde{c}(x) = \sum_{y \sim x} c_{xy}.$$

□

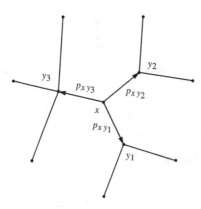

Figure 8.1: Transition probabilities p_{xy} at a vertex x (in V).

Define the transition operator

$$(Pf)(x) = \sum_{y \sim x} p_{xy} f(y),$$

then $P\mathbb{1} = \mathbb{1}$. Note that

$$\Delta f = 0 \Longleftrightarrow Pf = f.$$

Electrical Current as Frame Coefficients The role of the graph-network setting $(V, E, c, \mathcal{H}_E)$ introduced above is, in part, to model a family of electrical networks; one application among others. Here G is a graph with vertices V, and edges E. Since we study large networks, it is helpful to take V infinite, but countable. Think of a network of resistors placed on the edges in G. In this setting, the functions $v_{(x,y)}$ in $\mathcal{H}_E$, indexed by pairs of vertices, represent dipoles. They measure voltage drop in the network through all possible paths between the two vertices x and y. Now the conduction function c is given, and so (electrical) current equals the product of conductance and voltage drop; in this case voltage drop is computed over the paths made up of edges from x to y. For infinite systems (V, E, c) the corresponding dipoles v_{xy} are *not* in $l^2(V)$, but they are always in $\mathcal{H}_E$; see Lemma 8.7 below.

For a fixed function u in $\mathcal{H}_E$ (voltage), and e in E we calculate the current $I(u, e)$, and we show that these numbers yield *frame coefficients* in a natural Parseval frame (for $\mathcal{H}_E$) where the frame vectors making up the Parseval frame are v_e, $e = (x, y)$ in E.

When a conductance function $c : E \to \mathbb{R}_+ \cup \{0\}$ is given, we consider the energy Hilbert space $\mathcal{H}_E$ (depending on c), with norm and inner product:

$$\langle u, v \rangle_{\mathcal{H}_E} := \frac{1}{2} \sum \sum_{(x,y) \in E} c_{xy} \left(\overline{u(x)} - \overline{u(y)} \right) (v(x) - v(y)), \text{ and} \quad (8.23)$$

$$\|u\|_{\mathcal{H}_E}^2 = \langle u, u \rangle_{\mathcal{H}_E} = \frac{1}{2} \sum \sum_{(x,y) \in E} c_{xy} |u(x) - u(y)|^2 < \infty. \quad (8.24)$$

We shall assume that (V, E, c) is *connected* (see Definition 8.3); and it is known that $\mathcal{H}_E$ is a Hilbert space of functions on V; functions defined modulo constants. Further, for any pair of vertices $x, y \in V$, there is a unique dipole vector $v_{xy} \in \mathcal{H}_E$ such that

$$\langle v_{xy}, u \rangle_{\mathcal{H}_E} = u(x) - u(y) \quad (8.25)$$

holds for all $u \in \mathcal{H}_E$, see Lemma 8.7 below.

Dipoles

Let $(V, E, c, \mathscr{H}_E)$ be as described above, and assume that (V, E, c) is connected. Note that "vectors" in $\mathscr{H}_E$ are equivalence classes of functions on V (= the vertex set in the graph $G = (V, E)$).

Lemma 8.7. *For every pair of vertices* $x, y \in V$, *there is a* <u>*unique*</u> *vector* $v_{xy} \in \mathscr{H}_E$ *satisfying*

$$\langle u, v_{xy} \rangle_{\mathscr{H}_E} = u(x) - u(y) \tag{8.26}$$

for all $u \in \mathscr{H}_E$. *Thus, we have a* <u>*relative*</u> *reproducing kernel Hilbert space.*

Proof. Fix a pair of vertices x, y as above, and pick a finite path of edges $(x_i, x_{i+1}) \in E$ such that $c_{x_i, x_{i+1}} > 0$, and $x_0 = y$, $x_n = x$. Then

$$u(x) - u(y) = \sum_{i=0}^{n-1} u(x_{i+1}) - u(x_i)$$

$$= \sum_{i=0}^{n-1} \frac{1}{\sqrt{c_{x_i x_{i+1}}}} \sqrt{c_{x_i x_{i+1}}} (u(x_{i+1}) - u(x_i)); \tag{8.27}$$

and, by Schwarz, we have the following estimate:

$$|u(x) - u(y)|^2 \leq \left(\sum_{i=0}^{n-1} \frac{1}{c_{x_i x_{i+1}}} \right) \sum_{j=0}^{n-1} c_{x_j x_{j+1}} |u(x_{j+1}) - u(x_j)|^2$$

$$\leq (\text{Const}_{xy}) \|u\|_{\mathscr{H}_E}^2,$$

valid for all $u \in \mathscr{H}_E$, where we used (8.24) in the last step of this *a priori* estimate. But this states that the linear functional:

$$L_{xy} : \mathscr{H}_E \ni u \longmapsto u(x) - u(y) \tag{8.28}$$

is continuous on $\mathscr{H}_E$ w.r.t. the norm $\|\cdot\|_{\mathscr{H}_E}$. Hence existence and uniqueness for $v_{xy} \in \mathscr{H}_E$ follows from Riesz' theorem (see, e.g., Lemma 1.18). We get $\exists! v_{xy} \in \mathscr{H}_E$ such that

$$L_{xy}(u) = \langle v_{xy}, u \rangle_{\mathscr{H}_E}, \quad \forall u \in \mathscr{H}_E. \qquad \square$$

Theorem 8.8. *Let* $(V, E, c, E^{(ori)})$ *and* $\mathscr{H}_E$ *be as above; then the system of vectors*

$$w_{xy} := \sqrt{c_{xy}} v_{xy}, \quad \text{indexed by } (xy) \in E^{(ori)} \tag{8.29}$$

is a Parseval frame for $\mathscr{H}_E$.

Proof. Indeed we have for $u \in \mathscr{H}_E$:

$$\|u\|_{\mathscr{H}_E}^2 \overset{\text{(by (8.24))}}{=} \sum_{(xy) \in E^{(ori)}} c_{xy} |u(x) - u(y)|^2$$

$$\overset{\text{(by (8.25))}}{=} \sum_{(xy) \in E^{(ori)}} c_{xy} \left| \langle v_{xy}, u \rangle_{\mathscr{H}_E} \right|^2$$

$$= \sum_{(xy) \in E^{(ori)}} \left| \langle \sqrt{c_{xy}} v_{xy}, u \rangle_{\mathscr{H}_E} \right|^2$$

$$\overset{\text{(by (8.29))}}{=} \sum_{(xy) \in E^{(ori)}} \left| \langle w_{xy}, u \rangle_{\mathscr{H}_E} \right|^2$$

which is the desired conclusion. □

Remark 8.9. While the vectors $w_{xy} := \sqrt{c_{xy}} v_{xy}$, $(xy) \in E^{(ori)}$, form a Parseval frame in $\mathscr{H}_E$ in the general case, typically this frame is not an orthogonal basis (ONB) in $\mathscr{H}_E$; although it is in some special cases.

Now, we may introduce the resistance metric

$$R(x, y) := \|v_x - v_y\|_{\mathscr{H}_E}^2,$$

and set

$$G(x, y) = \langle v_x, v_y \rangle_{\mathscr{H}_E} = \frac{R(x, o) + R(y, o) - R(x, y)}{2}.$$

It can be shown then that

$$\Delta_{(\cdot)} G(\cdot, x) = \delta_x.$$

That is, G is the Greens function for the Laplacian operator.

Example 8.10. In the classical case, we consider a domain $\Omega \subset \mathbb{R}^n$ and the associated energy space $\mathscr{H}_E$, with

$$\|f\|_{\mathscr{H}_E}^2 = \frac{1}{2} \int |\nabla f|^2.$$

Then $\Delta G(x, y) = \delta(x - y)$, where $\Delta = -\nabla^2$ is the standard Laplacian operator.

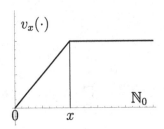

Figure 8.2: Kernel function for Brownian motion covariance (Discrete case).

Example 8.11 (Discrete Brownian motion). Let the vertex set be $V = \mathbb{N}_0 = \{0, 1, 2, \ldots\}$, with edges $E = \{(n, n+1)\}$ (see Figure 8.2). Set

$$\Delta f\,(n) = 2f\,(n) - f\,(n-1) - f(n+1).$$

The energy space is defined as

$$\mathscr{H}_E = \left\{ f : V \to \mathbb{C} \mid f\,(0) = 0,\ \|f\|_{\mathscr{H}_E} < \infty \right\}$$

where

$$\|f\|_{\mathscr{H}_E}^2 = \sum_n |f\,(n) - f\,(n+1)|^2.$$

Set

$$v_{ox}\,(y) = \begin{cases} y & \text{if } y < x \\ x & \text{if } y \geq x. \end{cases}$$

In this case, we have

$$\Delta v_x\,(y) = \delta_{xy},$$

$$\langle v_x, v_y \rangle_{\mathscr{H}_E} = v_x\,(y) - v_x\,(o) = \delta_{xy}.$$

Moreover, for all $f \in \mathscr{H}_E$, we have

$$\langle f, v_x \rangle = \sum_{n=1}^{x} (f\,(n) - f\,(n-1)) = f\,(x) - f\,(o) = f(x).$$

Let $(V, E, e, \mathscr{H}\,(= \mathscr{H}_E))$ be as above. By general theory of Minlos or Kolmogorov, there exists a probability space $(\Omega, \mathscr{C}, \mathbb{P})$, such that for $W_x := W_{v_x}$, we have

$$\mathbb{E}\,(W_x W_y) = \langle v_x, v_y \rangle_{\mathscr{H}} = G(x, y),$$

where G is the discrete Greens function. Note that in $\mathscr{H} = \mathscr{H}(V, E, c)$, the function δ_x has the following representation

$$\delta_x = c(x) v_x - \sum_{y \sim x} c_{xy} v_y.$$

Note that since $\#N_x < \infty$, the summation above is in fact a finite sum. Also,

$$\|\delta_x\|_{\mathscr{H}}^2 = c(x) = \sum_{y \sim x} c_{xy}.$$

8.2.1 Deciding when the Dirac point functions δ_x are in $\mathscr{H}(K)$

Discrete RKHS Let $S \times S \xrightarrow{K} \mathbb{R}$ be p.d., and let $\mathscr{H}(K)$ be the RKHS. Let $s \in S$ be fixed.

Question. *When is $\delta_s \in \mathscr{H}(K)$?*

Answer. Consider all finite subsets $F \subset S$, and let

$$K_F := K|_{F \times F}$$

as a finite $|F| \times |F|$ matrix. Then $\delta_s \in \mathscr{H}(K)$ if and only if

$$\sup_F K_F^{-1}(\delta_s)(s) < \infty,$$

where F runs over the filter of all finite subsets $\mathscr{F}(S)$ of S. Details below.

Lemma 8.12. *Let $F \in \mathscr{F}(S)$, $x_1 \in F$, then*

$$\left(K_F^{-1}\delta_{x_1}\right)(x_1) = \|P_F \delta_{x_1}\|_{\mathscr{H}}^2. \tag{8.30}$$

Proof. Setting $\zeta^{(F)} := K_F^{-1}(\delta_{x_1})$, we have

$$P_F(\delta_{x_1}) = \sum_{y \in F} \zeta^{(F)}(y) K_F(\cdot, y)$$

and for all $z \in F$,

$$\underbrace{\sum_{z \in F} \zeta^{(F)}(z) P_F(\delta_{x_1})(z)}_{\zeta^{(F)}(x_1)} = \sum_F \sum_F \zeta^{(F)}(z) \zeta^{(F)}(y) K_F(z, y) \tag{8.31}$$

$$= \|P_F \delta_{x_1}\|_{\mathscr{H}}^2.$$

Note that the LHS of (8.31) is given by

$$\|P_F \delta_{x_1}\|^2_{\mathscr{H}} = \langle P_F \delta_{x_1}, \delta_{x_1} \rangle_{\mathscr{H}}$$

$$= \sum_{y \in F} \left(K_F^{-1} \delta_{x_1} \right) (y) \langle K_y, \delta_{x_1} \rangle_{\mathscr{H}}$$

$$= \left(K_F^{-1} \delta_{x_1} \right) (x_1) = K_F^{-1}(x_1, x_1).$$

$\square$

Corollary 8.13. *If* $\delta_{x_1} \in \mathscr{H}$, *then*

$$\sup_{F \in \mathscr{F}(S)} \left(K_F^{-1} \delta_{x_1} \right) (x_1) = \|\delta_{x_1}\|^2_{\mathscr{H}}. \tag{8.32}$$

Theorem 8.14. *Given* S, $K : S \times S \to \mathbb{R}$ *positive definite (p.d.). Let* $\mathscr{H} = \mathscr{H}(K)$ *be the corresponding RKHS. Assume* S *is countable and infinite. Then the following three conditions (i)–(iii) are equivalent;* $x_1 \in S$ *is fixed:*

(i) $\delta_{x_1} \in \mathscr{H}$;
(ii) $\exists C_{x_1} < \infty$ *such that for all* $F \in \mathscr{F}(S)$, *the following estimate holds:*

$$|\xi(x_1)|^2 \leq C_{x_1} \sum_{F \times F} \sum \overline{\xi(x)} \xi(y) K(x, y) \tag{8.33}$$

(iii) *For* $F \in \mathscr{F}(S)$, *set*

$$K_F = (K(x, y))_{(x,y) \in F \times F} \tag{8.34}$$

as a $\#F \times \#F$ *matrix. Then*

$$\sup_{F \in \mathscr{F}(S)} \left(K_F^{-1} \delta_{x_1} \right) (x_1) < \infty. \tag{8.35}$$

Proof. (i)$\Rightarrow$(ii) For $\xi \in l^2(F)$, set

$$h_\xi = \sum_{y \in F} \xi(y) K_y(\cdot) \in \mathscr{H}_F.$$

Then $\langle \delta_{x_1}, h_\xi \rangle_{\mathscr{H}} = \xi(x_1)$ for all ξ.
    Since $\delta_{x_1} \in \mathscr{H}$, then by Schwarz:

$$\left| \langle \delta_{x_1}, h_\xi \rangle_{\mathscr{H}} \right|^2 \leq \|\delta_{x_1}\|^2_{\mathscr{H}} \sum_{F \times F} \sum \overline{\xi(x)} \xi(y) K(x, y). \tag{8.36}$$

But $\langle \delta_{x_1}, K_y \rangle_{\mathscr{H}} = \delta_{x_1,y} = \begin{cases} 1 & y = x_1 \\ 0 & y \neq x_1 \end{cases}$; hence $\langle \delta_{x_1}, h_\xi \rangle_{\mathscr{H}} = \xi(x_1)$, and so (8.36) implies (8.33).

(ii)$\Rightarrow$(iii) Recall the matrix

$$K_F := (\langle K_x, K_y \rangle)_{(x,y) \in F \times F}$$

as a linear operator $l^2(F) \to l^2(F)$, where

$$(K_F \varphi)(x) = \sum_{y \in F} K_F(x,y)\, \varphi(y), \quad \varphi \in l^2(F). \tag{8.37}$$

By (8.33), we have

$$\ker(K_F) \subset \{\varphi \in l^2(F) : \varphi(x_1) = 0\}. \tag{8.38}$$

Equivalently,

$$\ker(K_F) \subset \{\delta_{x_1}\}^\perp \tag{8.39}$$

and so $\delta_{x_1}|_F \in \ker(K_F)^\perp = \operatorname{ran}(K_F)$, and $\exists\, \zeta^{(F)} \in l^2(F)$ s.t.

$$\delta_{x_1}\Big|_F = \underbrace{\sum_{y \in F} \zeta^{(F)}(y)\, K(\cdot, y)}_{=:h_F}. \tag{8.40}$$

Claim. $P_F(\delta_{x_1}) = h_F$, where $P_F =$ projection onto $\mathscr{H}_F$. Indeed, we only need to prove that $\delta_{x_1} - h_F \in \mathscr{H} \ominus \mathscr{H}_F$, i.e.,

$$\langle \delta_{x_1} - h_F, K_z \rangle_{\mathscr{H}} = 0, \quad \forall z \in F. \tag{8.41}$$

But, by (8.40),

$$\mathrm{LHS}_{(8.41)} = \delta_{x_1,z} - \sum_{y \in F} K(z,y)\, \zeta^{(F)}(y) = 0.$$

This proves the claim.

If $F \subset F'$, $F, F' \in \mathscr{F}(S)$, then $\mathscr{H}_F \subset \mathscr{H}_{F'}$, and $P_F P_{F'} = P_F$ by easy facts for projections. Hence

$$\|P_F \delta_{x_1}\|_{\mathscr{H}}^2 \leq \|P_{F'} \delta_{x_1}\|_{\mathscr{H}}^2, \quad h_F := P_F(\delta_{x_1})$$

and

$$\lim_{F \nearrow S} \|\delta_{x_1} - h_F\|_{\mathscr{H}} = 0.$$

(iii)$\Rightarrow$(i) Follows from Lemma 8.12 and Corollary 8.13. $\qquad\square$

8.3   Guide to the literature

While we have cited some paper from the Bibliography in our discussion inside the chapter, readers not familiar with the ideas involved may wish to consult the following sources: [Bog98, BZ12, Che13, CP82, JT18a, JT18b, Ok08, Sch40, vNS41, Hid80].

Appendix A

Some Biographical Sketches

Inside the book, the following authors are cited frequently N. Aronszajn, V. Bargmann, S. Bergman, M. Born, L. Carleson, P. Dirac, W. Doeblin, J. Doob, V. Fock, I. Gelfand, W. Heisenberg, D. Hilbert, K. Itô, A. Kolmogorov, G. Mackey, P. Malliavin, A. Markov, E. Nelson, M-A Parseval, I. Schoenberg, H.A. Schwarz, L. Schwartz, J. Schwartz, I. Segal, M. Stone, J. von Neumann, N. Wiener, and A. Wightman. Below a short bio for each of them.

Nachman Aronszajn (1907–1980) Polish American mathematician. While he made contributions to diverse areas of mathematical analysis (PDE, potential theory, boundary value problems), he is best known for his pioneering paper on reproducing kernel Hilbert space (RKHS), and their applications, [Aro50]. He is especially well known for what now goes by the name, the Moore–Aronszajn theorem. He held Ph.D.s from both the University of Warsaw, (1930, Stefan Mazurkiewicz), and from Paris University, (1935; Maurice Fréchet). His career from 1951 was spent at University of Kansas.

Valentine "Valya" Bargmann (1908–1989) German-American mathematical physicist. PhD from the University of Zürich, under Gregor Wentzel. Worked at the Institute for Advanced Study in Princeton (1937–46) as an assistant to Albert Einstein. Was Professor at Princeton University since 1946, in the Mathematics and Physics departments. He found the irreducible unitary representations of $SL_2(\mathbb{R})$ and of the Lorentz group (1947). And he found the corresponding holomorphic representation realized in Segal–Bargmann space (1961). The latter is a reproducing kernel

Hilbert space (RKHS), with associated kernel, now called the Bargmann kernel, (— different from the Bergman kernel which is named after Stefan Bergman. It refers to a quite different RKHS, the Bergman space.) With Eugene Wigner, Valya Bargmann found what is now called the Bargmann–Wigner equations for particles of arbitrary spin; and he discovered the Bargmann–Michel–Telegdi equation, describing relativistic precession of the maximum number of bound states of quantum mechanical potentials (1952).

Stefan Bergman (1895–1977) Polish-born American mathematician. He worked on diverse areas in analytic functions on domains in several complex variables; and he is perhaps best known for the kernel function named after him. It is known today as the Bergman kernel. Bergman taught at Stanford University for most of his life. His Ph.D. is from Berlin University (1921, advisor, Richard von Mises) with a dissertation in Fourier analysis.

Max Born (1882–1970) A German physicist and mathematician, a pioneer in the early development of quantum mechanics; also in solid-state physics, and optics. Won the 1954 Nobel Prize in Physics for his "fundamental research in Quantum Mechanics, especially in the statistical interpretation of the wave function." His assistants at Göttingen, between the two World Wars, included Enrico Fermi, Werner Heisenberg, and Eugene Wigner, among others. His early education was at Breslau, where his fellow students included Otto Toeplitz and Ernst Hellinger.

In 1926, he formulated the now-standard interpretation of the probability density function for states (represented as equivalence classes of solutions to the Schrödinger equation.) After the Nazi Party came to power in Germany in 1933, Born was suspended. Subsequently he held positions at Johns Hopkins University, at Princeton University, and he settled down at St John's College, Cambridge (UK). A quote: "I believe that ideas such as absolute certitude, absolute exactness, final truth, etc. are figments of the imagination which should not be admissible in any field of science. On the other hand, any assertion of probability is either right or wrong from the standpoint of the theory on which it is based." Max Born (1954.)

Lennart Axel Edvard Carleson (1928–) A Swedish mathematician, pioneer and a leader in the area of harmonic analysis; is perhaps best known for his proof of Lusin's conjecture: Carleson proved the almost everywhere convergence of Fourier series for square-integrable functions (now known as Carleson's theorem). He was a student of Arne Beurling (Ph.D. from Uppsala University in 1950). He held a post-doc at Harvard University. Through his career he held long term positions at Uppsala University, the Royal Institute of Technology in Stockholm, and at the University of California, Los Angeles. He also worked in the theory of complex dynamical systems. With Benedicks, he proved existence of strange attractors in Hénon map in 1991.

His awards include the Wolf Prize in Mathematics in 1992, the Lomonosov Gold Medal in 2002, the Sylvester Medal in 2003, and the Abel Prize in 2006.

Paul Adrien Maurice Dirac (1902–1984) Cited in connection with the "Dirac equation" and especially our notation for vectors and operators in Hilbert space, as well as the axioms of observables, states and measurements. P. Dirac, an English theoretical physicist; fundamental contributions to the early development of both quantum mechanics and quantum electrodynamics. He was the Lucasian Professor of Mathematics at the University of Cambridge. Notable discoveries, the Dirac equation, which describes the behavior of fermions and predicted the existence of antimatter. Dirac shared the Nobel Prize in Physics for 1933 with Erwin Schrödinger, "for the discovery of new productive forms of atomic theory." A rare interview with Dirac; see the link: http://www.math.rutgers.edu/~greenfie/mill_courses/math421/int.html

Wolfgang Doeblin (Vincent Doblin 1915–1940) A French-German mathematician. Is the son of a Jewish-German novelist, Alfred Döblin, and Erna Reiss. His family escaped from Nazi Germany to France where he became a French citizen. As a soldier in the French army in 1939, he was quartered at Givet, in the Ardennes. While stationed, he wrote down his last work covering stochastic calculus, and inspired by the Chapman–Kolmogorov equation. Just before his tragic death, he sent it as a "pli cacheté" (sealed envelope) to the French Academy of Sciences. In May 1940, his company was caught in the German attack. And on 21 June, Doeblin shot himself in Housseras when German troops came within sight. In his last moments, he burned his mathematical notes. However, the sealed envelope was opened in 2000, revealing that Doeblin had obtained major results on stochastic calculusand diffusion processes, ahead of its time, and had anticipated many subsequent advances in the field. Some authors now use the name Doeblin's lemma for Itô's lemma; others say Doeblin-Itô.

Joseph Leo "Joe" Doob (1910–2004) An American mathematician, known for pioneering work in harmonic analysis and in probability theory, especially in stochastic processes. His foundational and pioneering work on probability and stochastic processes includes martingales, Markov processes, and stationary processes. His book *Stochastic Processes*, published in 1953, became one of the most influential books in modern probability theory. His later book, *Classical Potential Theory and Its Probabilistic Counterpart*, was also influential, and it served to bring together tools from analysis and probability.

Vladimir Aleksandrovich Fock (or Fok, 1898–1974) Was a Soviet physicist. He is best known for his foundational work in quantum mechanics and quantum electrodynamics. He was one of the pioneers in the development of quantum physics, and the theory of gravitation. His other contributions include theoretical optics, physics of continuous media, the Klein–Gordon equation. His name is used for such fundamental concepts as Fock space, the Fock representation and Fock state, and the Hartree–Fock method.

Israel Moiseevich Gelfand (1913–2009) Is the "G" in GNS (Gelfand–Naimark–Segal), the correspondence between states and cyclic representations.

I. Gelfand, also written Israïl Moyseyovich Gel'fand, or Izrail M. Gelfand, a Russian-American mathematician; major contributions to many branches of mathematics: representation theory and functional analysis. The recipient of numerous awards and honors, including the Order of Lenin and the Wolf Prize — a lifelong academic, serving decades as a professor at Moscow State University and, after immigrating to the United States shortly before his 76th birthday, at Rutgers University.

Werner Karl Heisenberg (1901–1976) Is the Heisenberg of the Heisenberg uncertainty principle for the operators P (momentum) and Q (position), and of matrix mechanics, as the first mathematical formulation of quantum observables. In Heisenberg's picture, the dynamics, the observables are studied as function of time; by contrast to Schrödinger's model which have the states (wavefunctions) functions of time, and satisfying a PDE wave equation, now called the Schrödinger equation. In the late 1920's, the two pictures, that of Heisenberg and of Schrödinger were thought to be irreconcilable. Work of von Neumann in 1932 demonstrated that they in fact are equivalent.

W. Heisenberg; one of the key creators of quantum mechanics. A 1925 paper was a breakthrough. In the subsequent series of papers with Max Born and Pascual Jordan, this matrix formulation of quantum mechanics took a mathematical rigorous formulation. In 1927 he published his uncertainty principle. Heisenberg was awarded the Nobel Prize in Physics for 1932 "for the creation of quantum mechanics." He made important contributions to the theories of the hydrodynamics of turbulent flows, the atomic nucleus.

David Hilbert (1862–1943) Cited in connection with the early formulations of the theory of operators in (what is now called) Hilbert space. The name Hilbert space was suggested by von Neumann who studied with Hilbert in the early 1930s, before he moved to the USA. (The early papers by von Neumann are in German.)

D. Hilbert is recognized as one of the most influential and universal mathematicians of the 19th and early 20th centuries. Discovered and developed invariant theory and axiomatization of geometry. In his 1900 presentation of a collection of research problems, he set the course for much of the mathematical research of the 20th century.

Kiyoshi Itô (1915–2008) Cited in connection with Brownian motion, Itô-calculus, and stochastic processes. Making connection to functional analysis via the theory of semigroups of operators (E. Hille, Kōsaku Yosida, and R.S. Phillips). The semigroup theorem is also known by the name, the Feller–Miyadera–Phillips theorem. Itô's career spanned diverse corners of the World, Japan, Stanford University, Aarhus University, Denmark (with Yosida); Cornell University, and the IAS Princeton. His list of awards includes the inaugural Gauss Prize in 2006 by the International Mathematical Union, and Japan's Order of Culture.

Andrey Nikolaevich Kolmogorov (1903–1987) Russian (Soviet at the time). Kolmogorov was a leading mathematician of the 20th Century, with pioneering contributions in analysis, probability theory, turbulence, information theory, and in computational complexity. A number of central notions in mathematics are named after him. The following list covers those of relevance to the topics in the present book, but it is only a fraction of the full list: Kolmogorov's axioms for probability theory, the Kolmogorov–Fokker–Planck equations, the Chapman–Kolmogorov equation, the Kolmogorov–Arnold–Moser (KAM) theorem, Kolmogorov's extension theorem, Kolmogorov complexity, and Kolmogorov's zero-one law.

(Most of Kolmogorov's work was done during the Cold War, and the scientists on either side of the Iron Curtain at the time had very little access to parallel developments on the other side.)

George Whitelaw Mackey (1916–2006) The first "M" in the "Mackey-machine," a systematic tool for constructing unitary representations of Lie groups, as induced representations. A pioneer in non-commutative harmonic analysis, and its applications to physics, to number theory, and to ergodic theory.

Paul Malliavin (1925–2010) A French mathematician. His research covered harmonic analysis, functional analysis, and stochastic processes. Is perhaps best known for what is now called Malliavin calculus, an infinite-dimensional variational calculus for stochastic processes. Applications include a probabilistic proof of the existence and smoothness of densities for the solutions to certain stochastic differential equations. Malliavin's calculus applies to, for example, random variables arising in computation of financial derivatives (financial products created from underlying securities), to the Clark-Ocone formula for martingale representations, and to stochastic filtering.

Andrey Andreyevich Markov (also spelled Markoff) (1856–1922) A Russian mathematician, known for his work on stochastic processes, and especially for what we now call Markov chains (discrete "time"), and Markov processes (in short, memory-free transition processes.) Both have applications to a host of statistical models. Loosely speaking, a Markov process is 'memoryless': the future of the process depends only on its present state, as opposed to the process's full history. So, conditioned on the present state, future and past are independent. With his younger brother Vladimir Andreevich Markov, he proved what is now called the Markov brothers' inequality. His son, Andrei Andreevich Markov (1903–1979), made important contributions to recursive function theory.

Edward (Ed) Nelson (1932–2014) Cited in connection with spectral representation, and Brownian motion.

Marc-Antoine Parseval (1755–1836) A French mathematician, known for what is now called the Parseval's theorem. It has become truly foundational, leading to effective use of Hilbert space methods in harmonic analysis, e.g., in making precise the unitarity property of the Fourier transform, and of related transforms in analysis. It has led to the now flourishing theory of frame analysis in a much wider context of pure and applied mathematics.

Isaac Jacob Schoenberg (1903–1990) A Romanian-American mathematician, perhaps best known for his discovery of splines. Conditionally negative definite functions, and their use in metric geometry, are also due to him. Also known for his influential work on total positivity, and on variation-diminishing linear transformations. A totally positive matrix is a square matrix for which all square submatrices have non-negative determinant.

Karl Hermann Amandus Schwarz (1843–1921) Is the Schwarz of the Cauchy–Schwarz inequality. H.A. Schwarz is German and is a contemporary of K. Weierstrass. H.A. Schwarz, a German mathematician, known for his work in complex analysis. At Göttingen, he pioneered of function theory, differential geometry and the calculus of variations.

Laurent-Moïse Schwartz (1915–2002) Is the Schwartz (French) of the theory of distributions (dating the 1950s), also now named "generalized functions" in the books by Gelfand *et al.* Parts of this theory were developed independently on the two sides of the Iron-Curtain; — in the time of the Cold War.

Jacob Theodore "Jack" Schwartz (1930–2009) Is the Schwartz of the book set "linear operators" by Dunford and Schwartz. Vol II is one of the best presentation of the theory of unbounded operators.

Irving Ezra Segal (1918–1998) Cited in connection with the foundations of functional analysis, and pioneering research in mathematical physics. Is the "S" in GNS (Gelfand–Naimark–Segal). Segal proved the Plancherel theorem in a very general framework: locally compact unimodular groups. For any locally compact unimodular group, Segal established a Plancherel formula. Segal showed that there is a Plancherel formula, despite the fact that it may not be feasible, for all locally compact unimodular groups, to "write down" all the irreducible unitary representations.

Marshall Harvey Stone (1903–1989) Is the "S" in the Stone–Weierstrass theorem; and in the Stone–von Neumann uniqueness theorem; the latter to the effect that any two representations of Heisenberg's commutation relations in the same (finite!) number of degrees of freedom are unitarily equivalent. Stone was the son of Harlan Fiske Stone, Chief Justice of the United States in 1941–1946. Marshall Stone completed a Harvard Ph.D. in 1926, with a thesis supervised by George David Birkhoff. He taught at Harvard, Yale, and Columbia University. And he was promoted to a full Professor at Harvard in 1937. In 1946, he became the chairman of the Mathematics Department at the University of Chicago. His 1932 monograph titled "Linear transformations in Hilbert space and their applications to analysis" develops the theory of selfadjoint operators, turning it into a form which is now a central part of functional analysis. Theorems that carry his name: The Banach–Stone theorem, The Glivenko–Stone theorem, Stone duality, The Stone–Weierstrass theorem, Stone's representation theorem for Boolean algebras, Stone's theorem for one-parameter unitary groups, Stone-Čech compactification, and The Stone-von Neumann uniqueness theorem. M. Stone and von Neumann are the two pioneers who worked at the same period. They were born at about the same time. Stone died at 1980's, and von Neumann died in the 1950's. For a bio and research overview.

John von Neumann (1903–1957) Cited in connection with the Stone–von Neumann uniqueness theorem, the deficiency indices which determine parameters for possible selfadjoint extensions of given Hermitian (formally selfadjoint, symmetric) with dense domain in Hilbert space.

J. von Neumann, Hungarian-American; inventor and polymath. He made major contributions to: foundations of mathematics, functional analysis, ergodic theory, numerical analysis, physics (quantum mechanics, hydrodynamics, and economics (game theory), computing (von Neumann architecture, linear programming, self-replicating machines, stochastic computing (Monte-Carlo[1])), — was a pioneer of the application of operator theory to quantum mechanics, a principal member of the Manhattan Project and the Institute for Advanced Study in Princeton. — A key figure in the development of game theory, cellular automata, and the digital computer.

Norbert Wiener (1894–1964) Cited in connection with Brownian motion, Wiener measure, and stochastic processes. And more directly, the

---

[1] "Monte-Carlo" means "simulation" with computer generated random number.

"Wiener" of Paley–Wiener spaces; — at the crossroads of harmonic analysis and functional analysis. Also the Wiener of filters in signal processing; high-pass/low-pass etc.

Arthur Strong Wightman (1922–2013) the "W" in the Wightman axioms for quantum field theory (also called the Gårding–Wightman axioms). Wightman was a leading mathematical physicist, a 1949 Princeton-PhD, and later a Professor at Princeton. He was one of the founders of quantum field theory. His former PhD students include Arthur Jaffe, Robert T. Powers, and Alan Sokal. The "PCT" in the title of refers to the combined symmetry of a quantum field theory under P for parity, C charge, and T time. "Spin and statistics" refers to a theorem to the effect that spin 1/2 particles obey Fermi–Dirac statistics, whereas integer spin 0, 1, 2 particles obey Bose–Einstein statistics. The Wightman axioms provide a basis for a mathematically rigorous perturbative approach to quantum fields. One of the Millennium Problems asks for a realization of the axioms in the case of Yang–Mills fields. A fundamental idea behind the axioms is that there should be a Hilbert space upon which the Poincaré group (of relativistic space-time) acts as a unitary representation. With this we get energy, momentum, angular momentum and center of mass (corresponding to boosts) realized as selfadjoint operators (unbounded) in this Hilbert space. A stability assumption places a restriction on the spectrum of the four-momentum, that it falls in the positive light cone (and its boundary). Then quantum fields entail realizations in the form of covariant representations of the Poincaré group. Further, the Wightman axioms entail operator valued distributions; in physics lingo, a smearing over Schwartz test functions. Here "operator-valued" refers to operators which are both non-commuting and unbounded. Hence the necessity for common dense domains; for example Gårding domains. The causal structure of the theory entails imposing either commutativity (CCR), or anticommutativity (CAR) rules for spacelike separated fields; and it further postulates the existence of a Poincaré-invariant and cyclic state, called the vacuum.

Bibliography

[AAH69] R. D. Adams, N. Aronszajn, and M. S. Hanna, Theory of Bessel potentials. Part III. Potentials on regular manifolds, *Ann. Inst. Fourier (Grenoble)* **19** (1969), no. 2, 279–338 (1970).

[AESW51] M. Aissen, A. Edrei, I. J. Schoenberg, and A. Whitney, On the generating functions of totally positive sequences, *Proc. Nat. Acad. Sci. U.S.A.* **37** (1951), 303–307.

[AFMP94] G. T. Adams, J. Froelich, P. J. McGuire, and V. I. Paulsen, Analytic reproducing kernels and factorization, *Indiana Univ. Math. J.* **43** (1994), no. 3, 839–856.

[AH84] D. B. Applebaum and R. L. Hudson, Fermion Itô's formula and stochastic evolutions, *Comm. Math. Phys.* **96** (1984), no. 4, 473–496.

[AJ12] D. Alpay and P. E. T. Jorgensen, Stochastic processes induced by singular operators, *Numer. Funct. Anal. Optim.* **33** (2012), no. 7–9, 708–735.

[AJ15] D. Alpay and P. Jorgensen, Spectral theory for Gaussian processes: reproducing kernels, boundaries, and L^2-wavelet generators with fractional scales, *Numer. Funct. Anal. Optim.* **36** (2015), no. 10, 1239–1285.

[AJL11] D. Alpay, P. Jorgensen and D. Levanony, A class of Gaussian processes with fractional spectral measures, *J. Funct. Anal.* **261** (2011), no. 2, 507–541.

[AJLM15] D. Alpay, P. Jorgensen, I. Lewkowicz and I. Martziano, Infinite product representations for kernels and iterations of functions, Recent advances in inverse scattering, Schur analysis and stochastic processes, *Oper. Theory Adv. Appl.*, vol. 244, Birkhäuser/Springer, Cham, 2015, pp. 67–87.

[AJS14] D. Alpay, P. Jorgensen and G. Salomon, On free stochastic processes and their derivatives, *Stochastic Process. Appl.* **124** (2014), no. 10, 3392–3411.

[AK18]   D. Auckly and P. Kuchment, On Parseval frames of exponentially decaying composite Wannier functions, Mathematical problems in quantum physics, Contemp. Math., vol. 717, *Amer. Math. Soc.*, Providence, RI, 2018, pp. 227–240.

[AKL16]   A. Andersson, R. Kruse and S. Larsson, *Duality in refined Sobolev–Malliavin spaces and weak approximation of SPDE*, Stoch. Partial Differ. Equ. Anal. Comput. **4** (2016), no. 1, 113–149.

[Alp01]   D. Alpay, *The Schur algorithm, reproducing kernel spaces and system theory*, SMF/AMS Texts and Monographs, vol. 5, American Mathematical Society, Providence, RI; Société Mathématique de France, Paris, 2001, Translated from the 1998 French original by Stephen S. Wilson.

[AØ15]   N. Agram and B. Øksendal, Malliavin calculus and optimal control of stochastic Volterra equations, *J. Optim. Theory Appl.* **167** (2015), no. 3, 1070–1094.

[App09]   D. Applebaum, Lévy processes and stochastic calculus, second ed., Cambridge Studies in Advanced Mathematics, vol. 116, Cambridge University Press, Cambridge, 2009.

[Aro50]   N. Aronszajn, Theory of reproducing kernels, *Trans. Amer. Math. Soc.* **68** (1950), 337–404.

[Aro61]   N. Aronszajn, Quadratic forms on vector spaces, Proc. Internat. Sympos. Linear Spaces (Jerusalem, 1960), Jerusalem Academic Press, Jerusalem; Pergamon, Oxford, 1961, pp. 29–87.

[Arv76a]   W. Arveson, Aspectral theorem for nonlinear operators, *Bull. Amer. Math. Soc.* **82** (1976), no. 3, 511–513.

[Arv76b]   W. Arveson, *Spectral theory for nonlinear random processes*, Symposia Mathematica, Vol. XX (Convegno sulle Algebre C^* e loro Applicazioni in Fisica Teorica, Convegno sulla Teoria degli Operatori Indice e Teoria K, INDAM, Rome, 1975), Academic Press, London, 1976, pp. 531–537.

[AS57]   N. Aronszajn and K. T. Smith, Characterization of positive reproducing kernels. Applications to Green's functions, *Amer. J. Math.* **79** (1957), 611–622.

[AW63]   H. Araki and E. J. Woods, Representations of the canonical commutation relations describing a nonrelativistic infinite free Bose gas, *J. Mathematical Phys.* **4** (1963), 637–662.

[AW73]   H. Araki and E. J. Woods, Topologies induced by representations of the canonical commutation relations, *Rep. Mathematical Phys.* **4** (1973), 227–254.

[Ban97]   W. Banaszczyk, The Minlos lemma for positive-definite functions on additive subgroups of $\mathbf{R}^n$, *Studia Math.* **126** (1997), no. 1, 13–25.

[BCL11]   R. Balan, P. Casazza and Z. Landau, Redundancy for localized frames, *Israel J. Math.* **185** (2011), 445–476.

[BCMS19]   M. Bownik, P. Casazza, A. W. Marcus and D. Speegle, Improved bounds in Weaver and Feichtinger conjectures, *J. Reine Angew. Math.* **749** (2019), 267–293.

[BCR84]   C. Berg, J. P. R. Christensen and P. Ressel, *Harmonic analysis on semigroups*, Graduate Texts in Mathematics, vol. 100, Springer-Verlag, New York, 1984, Theory of positive definite and related functions.

[BEGJ89a] O. Bratteli, G. A. Elliott, F. M. Goodman and P. E. T. Jorgensen, On Lie algebras of operators, *J. Funct. Anal.* **86** (1989), no. 2, 341–359.

[BEGJ89b] O. Bratteli, G. A. Elliott, F. M. Goodman and P. E. T. Jorgensen, Smooth Lie group actions on noncommutative tori, *Nonlinearity* **2** (1989), no. 2, 271–286.

[BEJ84]   O. Bratteli, G. A. Elliott and P. E. T. Jorgensen, Decomposition of unbounded derivations into invariant and approximately inner parts, *J. Reine Angew. Math.* **346** (1984), 166–193.

[Bel06]   D. R. Bell, The Malliavin calculus, Dover Publications, Inc., Mineola, NY, 2006, Reprint of the 1987 edition.

[BHS05]   M. Barnsley, J. Hutchinson and Ö. Stenflo, A fractal valued random iteration algorithm and fractal hierarchy, Fractals **13** (2005), no. 2, 111–146.

[BJ89]   O. Bratteli and P. E. T. Jorgensen, Conservative derivations and dissipative Laplacians, *J. Funct. Anal.* **82** (1989), no. 2, 404–411.

[BJKR84]   O. Bratteli, P. E. T. Jorgensen, A. Kishimoto and D. W. Robinson, A C^*-algebraic Schoenberg theorem, *Ann. Inst. Fourier (Grenoble)* **34** (1984), no. 3, 155–187.

[BJr82]   O. Bratteli and P. E. T. Jorgensen, Unbounded derivations tangential to compact groups of automorphisms, *J. Functional Analysis* **48** (1982), no. 1, 107–133.

[BJr83]   O. Bratteli and P. E. T. Jorgensen, Derivations commuting with abelian gauge actions on lattice systems, *Comm. Math. Phys.* **87** (1982/83), no. 3, 353–364.

[BLP19]   B. G. Bodmann, D. Labate and B. R. Pahari, Smooth projections and the construction of smooth Parseval frames of shearlets, *Adv. Comput. Math.* **45** (2019), no. 5–6, 3241–3264.

[Bog98]   V. I. Bogachev, *Gaussian measures*, Mathematical Surveys and Monographs, vol. 62, American Mathematical Society, Providence, RI, 1998.

[BØSW04]   F. Biagini, B. Øksendal, A. Sulem and N. Wallner, An introduction to white-noise theory and Malliavin calculus for fractional Brownian motion, *Proc. R. Soc. Lond. Ser. A Math. Phys. Eng. Sci.* **460** (2004), no. 2041, 347–372, Stochastic analysis with applications to mathematical finance.

[BQ15]   A. I. Bufetov and Y. Qiu, Equivalence of Palm measures for determinantal point processes associated with Hilbert spaces of holomorphic functions, *C. R. Math. Acad. Sci. Paris* **353** (2015), no. 6, 551–555.

[BR79]   O. Bratteli and D. W. Robinson, *Operator algebras and quantum statistical mechanics. Vol. 1*, Springer-Verlag, New York-Heidelberg,

1979, C^*- and W^*-algebras, algebras, symmetry groups, decomposition of states, Texts and Monographs in Physics.

[BR81]   O. Bratteli and D. W. Robinson, *Operator algebras and quantum-statistical mechanics. II*, Springer-Verlag, New York-Berlin, 1981, Equilibrium states. Models in quantum-statistical mechanics, Texts and Monographs in Physics.

[BSS89]   L. Blum, M. Shub and S. Smale, On a theory of computation and complexity over the real numbers: NP-completeness, recursive functions and universal machines, *Bull. Amer. Math. Soc. (N.S.)* **21** (1989), no. 1, 1–46.

[BSS90]   L. Blum, M. Shub and S. Smale, *On a theory of computation over the real numbers: NP completeness, recursive functions and universal machines [Bull. Amer. Math. Soc. (N.S.)* **21** *(1989), no. 1, 1–46; MR0974426 (90a:68022)]*, Workshop on Dynamical Systems (Trieste, 1988), Pitman *Res. Notes Math. Ser.*, vol. 221, Longman Sci. Tech., Harlow, 1990, pp. 23–52.

[BT14]   P. Broecker and S. Trebst, Rényi entropies of interacting fermions from determinantal quantum Monte Carlo simulations, *J. Stat. Mech. Theory Exp.* (2014), no. 8, P08015, 22.

[Buf16]   A. I. Bufetov, *Infinite determinantal measures and the ergodic decomposition of infinite Pickrell measures. II. Convergence of determinantal measures*, *Izv. Ross. Akad. Nauk Ser. Mat.* **80** (2016), no. 2, 16–32.

[BZ12]   I. Bilionis and N. Zabaras, Multi-output local Gaussian process regression: applications to uncertainty quantification, *J. Comput. Phys.* **231** (2012), no. 17, 5718–5746.

[CH13]   S. Chen and R. Hudson, Some properties of quantum Lévy area in Fock and non-Fock quantum stochastic calculus, *Probab. Math. Statist.* **33** (2013), no. 2, 425–434.

[Che13]   L. Chen, Generalized multiplicity-free representations of nongraded divergence-free Lie algebras, *J. Lie Theory* **23** (2013), no. 2, 507–549.

[CP82]   F. Coester and W. N. Polyzou, Relativistic quantum mechanics of particles with direct interactions, *Phys. Rev. D* (3) **26** (1982), no. 6, 1348–1367.

[DHK13]   B. K. Driver, B. C. Hall and T. Kemp, The large-N limit of the Segal-Bargmann transform on $\mathbb{U}_N$, *J. Funct. Anal.* **265** (2013), no. 11, 2585–2644.

[Dir35]   P. A. M. Dirac, The electron wave equation in de-Sitter space, *Ann. of Math.* (2) **36** (1935), no. 3, 657–669.

[Dir47]   P. A. M. Dirac, *The Principles of Quantum Mechanics*, Oxford, at the Clarendon Press, 1947, 3d ed.

[Dix77]   J. Dixmier, *Enveloping algebras*, North-Holland Publishing Co., Amsterdam-New York-Oxford, 1977, North-Holland Mathematical Library, Vol. 14, Translated from the French.

[Doo75]   J. L. Doob, Stochastic process measurability conditions, *Ann. Inst. Fourier (Grenoble)* **25** (1975), no. 3–4, xiv, 163–176.

[Doo89]   J. L. Doob, Kolmogorov's early work on convergence theory and foundations, *Ann. Probab.* **17** (1989), no. 3, 815–821.

[Doo96]   J. L. Doob, *The development of rigor in mathematical probability (1900–1950) [in development of mathematics 1900–1950 (luxembourg, 1992), 157–170, Birkhäuser, Basel, 1994; MR1298633 (95i:60001)]*, Amer. Math. Monthly **103** (1996), no. 7, 586–595.

[DS88]   N. Dunford and J. T. Schwartz, *Linear operators. Part II*, Wiley Classics Library, John Wiley & Sons, Inc., New York, 1988, Spectral theory. Selfadjoint operators in Hilbert space, With the assistance of William G. Bade and Robert G. Bartle, Reprint of the 1963 original, A Wiley-Interscience Publication.

[Fad18]   A. V. Fadeeva, *Parseval frames of serial shifts of a function in spaces of trigonometric polynomials*, Moscow Univ. Math. Bull. **73** (2018), no. 6, 239–244, Translation of Vestnik Moskov. *Univ. Ser. I Mat. Mekh.* **2**018, no. 6, 30–36.

[Fre14]   W. Freyn, Tame Fréchet structures for affine Kac-Moody groups, *Asian J. Math.* **18** (2014), no. 5, 885–928.

[GJ87]   J. Glimm and A. Jaffe, *Quantum physics*, second ed., Springer-Verlag, New York, 1987, A functional integral point of view.

[GK54]   B. V. Gnedenko and A. N. Kolmogorov, *Limit distributions for sums of independent random variables*, Addison-Wesley Publishing Company, Inc., Cambridge, Mass., 1954, Translated and annotated by K. L. Chung. With an Appendix by J. L. Doob.

[GP17]   S. Ghosh and Y. Peres, Rigidity and tolerance in point processes: Gaussian zeros and Ginibre eigenvalues, *Duke Math. J.* **166** (2017), no. 10, 1789–1858.

[Gro70]   L. Gross, *Abstract Wiener measure and infinite dimensional potential theory*, Lectures in Modern Analysis and Applications, II, Lecture Notes in Mathematics, Vol. 140. Springer, Berlin, 1970, pp. 84–116.

[Hei69]   W. Heisenberg, *Über quantentheoretische umdeutung kinematischer und mechanischer beziehungen.(1925) in: G*, Ludwig, Wellenmechanik, Einführung und Originaltexte, Akademie-Verlag, Berlin (1969), 195.

[Hid71]   T. Hida, Quadratic functionals of Brownian motion, *J. Multivariate Anal.* **1** (1971), no. 1, 58–69.

[Hid80]   T. Hida, *Brownian motion*, Applications of Mathematics, vol. 11, Springer-Verlag, New York, 1980, Translated from the Japanese by the author and T. P. Speed.

[Hid85]   T. Hida, Brownian motion and its functionals, *Ricerche Mat.* **34** (1985), no. 1, 183–222.

[Hid90]   T. Hida, Functionals of Brownian motion, Lectures in applied mathematics and informatics, Manchester Univ. Press, Manchester, 1990, pp. 286–329.

[Hid93]   T. Hida, *A role of the Lévy Laplacian in the causal calculus of generalized white noise functionals*, Stochastic processes, Springer, New York, 1993, pp. 131–139.

[Hid03]   T. Hida, *Laplacians in white noise analysis*, Finite and infinite dimensional analysis in honor of Leonard Gross (New Orleans, LA, 2001), Contemp. Math., vol. 317, Amer. Math. Soc., Providence, RI, 2003, pp. 137–142.

[Hid07]   T. Hida, *Infinite dimensional harmonic analysis from the viewpoint of white noise theory*, Harmonic, wavelet and p-adic analysis, World Sci. Publ., Hackensack, NJ, 2007, pp. 313–330.

[HKPS13]   T. Hida, H. H. Kuo, J. Potthoff and W. Streit, *White noise: An infinite dimensional calculus*, Mathematics and Its Applications, Springer Netherlands, 2013.

[HM08]   B. C. Hall and J. J. Mitchell, Isometry theorem for the Segal-Bargmann transform on a noncompact symmetric space of the complex type, *J. Funct. Anal.* **254** (2008), no. 6, 1575–1600.

[HP84]   R. L. Hudson and K. R. Parthasarathy, Quantum Ito's formula and stochastic evolutions, *Comm. Math. Phys.* **93** (1984), no. 3, 301–323.

[HPP00]   R. L. Hudson, K. R. Parthasarathy and S. Pulmannová, Method of formal power series in quantum stochastic calculus, *Infin. Dimens. Anal. Quantum Probab. Relat. Top.* **3** (2000), no. 3, 387–401.

[HS98]   T. Hida and S. Si, Innovations for random fields, *Infin. Dimens. Anal. Quantum Probab. Relat. Top.* **1** (1998), no. 4, 499–509.

[Hud14]   R. L. Hudson, Forward and backward adapted quantum stochastic calculus and double product integrals, *Russ. J. Math. Phys.* **21** (2014), no. 3, 348–361.

[Hut81]   J. E. Hutchinson, Fractals and self-similarity, *Indiana Univ. Math. J.* **30** (1981), no. 5, 713–747.

[Itô04]   K. Itô, *Stochastic processes*, Springer-Verlag, Berlin, 2004, Lectures given at Aarhus University, Reprint of the 1969 original, Edited and with a foreword by Ole E. Barndorff-Nielsen and Ken-iti Sato.

[Itô06]   K. Itô, *Essentials of stochastic processes*, Translations of Mathematical Monographs, vol. 231, American Mathematical Society, Providence, RI, 2006, Translated from the 1957 Japanese original by Yuji Ito.

[Jan97]   S. Janson, *Gaussian Hilbert spaces*, Cambridge Tracts in Mathematics, vol. 129, Cambridge University Press, Cambridge, 1997.

[Jør80]   P. E. T. Jorgensen, Unbounded operators: perturbations and commutativity problems, *J. Funct. Anal.* **39** (1980), no. 3, 281–307.

[JP91]   P. E. T. Jorgensen and R. T. Powers, Positive elements in the algebra of the quantum moment problem, Probab. Theory Related Fields **89** (1991), no. 2, 131–139.

[JP08]   P. E. T. Jorgensen and E. P. J. Pearse, *Operator theory of electrical resistance networks*, arXiv e-prints (2008), arXiv:0806.3881.

[JP10]   P. E. T. Jorgensen and E. P. J. Pearse, A Hilbert space approach to effective resistance metric, *Complex Anal. Oper. Theory* **4** (2010), no. 4, 975–1013.

[JP11a]   P. E. T. Jorgensen and E. P. J. Pearse, Gel′ fand triples and boundaries of infinite networks, *New York J. Math.* **17** (2011), 745–781.

[JP11b]   P. E. T. Jorgensen and E. P. J. Pearse, Resistance boundaries of infinite networks, Random walks, boundaries and spectra, *Progr. Probab.*, vol. 64, Birkhäuser/Springer Basel AG, Basel, 2011, pp. 111–142.

[JP11c]   P. E. T. Jorgensen and E. P. J. Pearse, Spectral reciprocity and matrix representations of unbounded operators, *J. Funct. Anal.* **261** (2011), no. 3, 749–776.

[JP13]   P. E. T. Jorgensen and E. P. J. Pearse, A discrete Gauss-Green identity for unbounded Laplace operators, and the transience of random walks, *Israel J. Math.* **196** (2013), no. 1, 113–160.

[JP14]   P. E. T. Jorgensen and E. P. J. Pearse, Spectral comparisons between networks with different conductance functions, *J. Operator Theory* **72** (2014), no. 1, 71–86.

[Jr68]   O. G. Jorsboe, *Equivalence or singularity of Gaussian measures on function spaces*, Various Publications Series, No. 4, Matematisk Institut, Aarhus Universitet, Aarhus, 1968.

[JT17]   P. Jorgensen and F. Tian, *Non-commutative analysis*, World Scientific Publishing Co. Pte. Ltd., Hackensack, NJ, 2017, With a foreword by Wayne Polyzou.

[JT18a]   P. Jorgensen and F. Tian, Infinite weighted graphs with bounded resistance metric, *Math. Scand.* **123** (2018), no. 1, 5–38.

[JT18b]   P. Jorgensen and F. Tian, Metric duality between positive definite kernels and boundary processes, *Int. J. Appl. Comput. Math.* **4** (2018), no. 1, Paper No. 3, 13.

[JT19a]   P. Jorgensen and F. Tian, On reproducing kernels, and analysis of measures, *Markov Process. Related Fields* **25** (2019), no. 3, 445–482.

[JT19b]   P. Jorgensen and F. Tian, Realizations and factorizations of positive definite kernels, *J. Theoret. Probab.* **32** (2019), no. 4, 1925–1942.

[Kü1]   T. Kühn, Covering numbers of Gaussian reproducing kernel Hilbert spaces, *J. Complexity* **27** (2011), no. 5, 489–499.

[Kat19]   V. Katsnelson, *On the completeness of Gaussians in a Hilbert functional space*, *Complex Anal. Oper. Theory* **13** (2019), no. 3, 637–658.

[Kol83]   A. N. Kolmogorov, *On logical foundations of probability theory*, Probability theory and mathematical statistics (Tbilisi, 1982), Lecture Notes in Math., vol. 1021, Springer, Berlin, 1983, pp. 1–5.

[Kra13]   S. G. Krantz, *Geometric analysis of the Bergman kernel and metric*, Graduate Texts in Mathematics, vol. 268, Springer, New York, 2013.

[KU87]   A. N. Kolmogorov and V. A. Uspensky, *Algorithms and randomness*, Proceedings of the 1st World Congress of the Bernoulli Society, Vol. 1 (Tashkent, 1986), VNU Sci. Press, Utrecht, 1987, pp. 3–53.

[KW17]   S. Krantz and P. M. Wójcicki, The weighted Bergman kernel and the Green's function, *Complex Anal. Oper. Theory* **11** (2017), no. 1, 217–225.

[L52]   P. Lévy, Convergence des séries aléatoires et loi normale, *C. R. Acad. Sci. Paris* **234** (1952), 2422–2424.

[L62]   P. Lévy, Le déterminisme de la fonction brownienne dans l'espace de Hilbert, *Ann. Sci. École Norm. Sup.* (3) **79** (1962), 377–398.

[L69]   P. Lévy, Une hiérarchie des probabilités plus ou moins nulles, application à certains nuages de points, *Enseign. Math.* (2) **15** (1969), 217–225.

[L92]   P. Lévy, *Processus stochastiques et mouvement brownien*, Les Grands Classiques Gauthier-Villars. [Gauthier-Villars Great Classics], Éditions Jacques Gabay, Sceaux, 1992, Followed by a note by M. Loève, Reprint of the second (1965) edition.

[Lax02]   P. D. Lax, *Functional analysis*, Pure and Applied Mathematics (New York), Wiley-Interscience [John Wiley & Sons], New York, 2002.

[Leb05]   H. Lebesgue, Sur le problème des aires, *Bull. Soc. Math. France* **33** (1905), 273–274.

[LG14]   G. Liu and X. Guo, Harish-Chandra modules over generalized Heisenberg-Virasoro algebras, *Israel J. Math.* **204** (2014), no. 1, 447–468.

[LLL11]   W. L. Liu, S. L. Lü and F. B. Liang, Kernel density discriminant method based on geodesic distance, *J. Fuzhou Univ. Nat. Sci. Ed.* **39** (2011), no. 6, 807–810, 818.

[LPW10]   H. Luschgy, G. Pagès and B. Wilbertz, Asymptotically optimal quantization schemes for Gaussian processes on Hilbert spaces, *ESAIM Probab. Stat.* **14** (2010), 93–116.

[LS08]   J. Li and Y. Su, Representations of the Schrödinger-Virasoro algebras, *J. Math. Phys.* **49** (2008), no. 5, 053512, 14.

[Lyo03]   R. Lyons, Determinantal probability measures, *Publ. Math. Inst. Hautes Études Sci.* (2003), no. 98, 167–212.

[Mac04]   G. W. Mackey, *Mathematical foundations of quantum mechanics*, Dover Publications, Inc., Mineola, NY, 2004, With a foreword by A. S. Wightman, Reprint of the 1963 original.

[Mal78]   P. Malliavin, *Stochastic calculus of variation and hypoelliptic operators*, Proceedings of the International Symposium on Stochastic Differential Equations (Res. Inst. Math. Sci., Kyoto Univ., Kyoto, 1976), Wiley, New York-Chichester-Brisbane, 1978, pp. 195–263.

[Nel58]   E. Nelson, Kernel functions and eigenfunction expansions, *Duke Math. J.* **25** (1958), 15–27.

[Nel69]   E. Nelson, *Topics in dynamics. I: Flows*, Mathematical Notes, Princeton University Press, Princeton, N.J.; University of Tokyo Press, Tokyo, 1969.

[Nua06]   D. Nualart, *The Malliavin calculus and related topics*, second ed., Probability and its Applications (New York), Springer-Verlag, Berlin, 2006.

[Ok08]   B. Øksendal, Stochastic partial differential equations driven by multi-parameter white noise of Lévy processes, *Quart. Appl. Math.* **66** (2008), no. 3, 521–537.

[Ols20]   G. Olshanski, Determinantal point processes and fermion quasifree states, *Commun. Math. Phys.* (2020).

[OR00]   H. Ouerdiane and A. Rezgui, Un théorème de Bochner-Minlos avec une condition d'intégrabilité, *Infin. Dimens. Anal. Quantum Probab. Relat. Top.* **3** (2000), no. 2, 297–302.

[PK88]   W. N. Polyzou and W. H. Klink, The structure of Poincaré covariant tensor operators in quantum mechanical models, *Ann. Physics* **185** (1988), no. 2, 369–400.

[Pol02]   W. N. Polyzou, Cluster properties in relativistic quantum mechanics of N-particle systems, *J. Math. Phys.* **43** (2002), no. 12, 6024–6063.

[Pow74]   R. T. Powers, Selfadjoint algebras of unbounded operators. II, *Trans. Amer. Math. Soc.* **187** (1974), 261–293.

[PP14]   R. Pemantle and Y. Peres, Concentration of Lipschitz functionals of determinantal and other strong Rayleigh measures, *Combin. Probab. Comput.* **23** (2014), no. 1, 140–160.

[PR14]   B. L. S. Prakasa Rao, Characterization of Gaussian distribution on a Hilbert space from samples of random size, *J. Multivariate Anal.* **132** (2014), 209–214.

[PR16]   V. I. Paulsen and M. Raghupathi, *An introduction to the theory of reproducing kernel Hilbert spaces*, Cambridge Studies in Advanced Mathematics, vol. 152, Cambridge University Press, Cambridge, 2016.

[Pri10]   N. Privault, Random Hermite polynomials and Girsanov identities on the Wiener space, *Infin. Dimens. Anal. Quantum Probab. Relat. Top.* **13** (2010), no. 4, 663–675.

[PS72a]   K. R. Parthasarathy and K. Schmidt, Factorisable representations of current groups and the Araki-Woods imbedding theorem, *Acta Math.* **128** (1972), no. 1–2, 53–71.

[PS72b]   K. R. Parthasarathy, *Positive definite kernels, continuous tensor products, and central limit theorems of probability theory*, Lecture Notes in Mathematics, Vol. 272, Springer-Verlag, Berlin-New York, 1972.

[PS11]   I. Penkov and K. Styrkas, *Tensor representations of classical locally finite Lie algebras*, Developments and trends in infinite-dimensional Lie theory, Progr. Math., vol. 288, Birkhäuser Boston, Inc., Boston, MA, 2011, pp. 127–150.

[PV05]   Y. Peres and B. Virág, Zeros of the i.i.d. Gaussian power series: a conformally invariant determinantal process, *Acta Math.* **194** (2005), no. 1, 1–35.

[RS75]   M. Reed and B. Simon, *Methods of modern mathematical physics. II. Fourier analysis, self-adjointness*, Academic Press [Harcourt Brace Jovanovich, Publishers], New York-London, 1975.

[Rud91]   W. Rudin, *Functional analysis*, second ed., International Series in
            Pure and Applied Mathematics, McGraw-Hill, Inc., New York, 1991.
[Sak98]   S. Sakai, *C*-algebras and W*-algebras*, Classics in Mathematics,
            Springer-Verlag, Berlin, 1998, Reprint of the 1971 edition.
[Sch32]   E. Schrödinger, Sur la théorie relativiste de l'électron et
            l'interprétation de la mécanique quantique, *Ann. Inst. H. Poincaré*
            2 (1932), no. 4, 269–310.
[Sch37]   I. J. Schoenberg, On certain metric spaces arising from Euclidean
            spaces by a change of metric and their imbedding in Hilbert space,
            Ann. of Math. (2) **38** (1937), no. 4, 787–793.
[Sch38]   I. J. Schoenberg, Metric spaces and completely monotone functions,
            Ann. of Math. (2) **39** (1938), no. 4, 811–841.
[Sch40]   E. Schrödinger, A method of determining quantum-mechanical
            eigenvalues and eigenfunctions, *Proc. Roy. Irish Acad. Sect. A.* **46**
            (1940), 9–16.
[Sch42]   I. J. Schoenberg, Positive definite functions on spheres, *Duke Math.
            J.* **9** (1942), 96–108.
[Sch66]   L. Schwartz, *Théorie des distributions*, Publications de l'Institut
            de Mathématique de l'Université de Strasbourg, No. IX-X. Nou-
            velle édition, entièrement corrigée, refondue et augmentée, Hermann,
            Paris, 1966.
[Sch68]   L. Schwartz, Réciproque du théorème de Sazonov-Minlos dans des
            cas non hilbertiens, *C. R. Acad. Sci. Paris Sér. A-B* **266** (1968),
            A7–A9.
[Sch94]   L. Schwartz, *Le mouvement brownien*, Fascicule de probabilités,
            Publ. Inst. Rech. Math. Rennes, vol. 1994, Univ. Rennes I, Rennes,
            1994, p. 25.
[Sch99]   E. Schrödinger, About Heisenberg uncertainty relation (original
            annotation by A. Angelow and M.-C. Batoni), *Bulgar. J. Phys.* **26**
            (1999), no. 5-6, 193–203 (2000), *Translation of Proc. Prussian Acad.
            Sci. Phys. Math. Sect.* **19** (1930), 296–303.
[Sha62]   D. Shale, Linear symmetries of free boson fields, *Trans. Amer. Math.
            Soc.* **103** (1962), 149–167.
[Šil47]   G. E. Šilov, On a property of rings of functions, *Doklady Akad. Nauk
            SSSR (N. S.)* **58** (1947), 985–988.
[SS64]   D. Shale and W. F. Stinespring, States of the Clifford algebra, *Ann.
            of Math.* (2) **80** (1964), 365–381.
[SS65]   D. Shale and W. F. Stinespring, Spinor representations of infinite
            orthogonal groups, *J. Math. Mech.* **14** (1965), 315–322.
[Sti55]   W. F. Stinespring, Positive functions on C*-algebras, *Proc. Amer.
            Math. Soc.* **6** (1955), 211–216.
[Sti74]   S. M. Stigler, Studies in the history of probability and statistics.
            XXXIII. Cauchy and the witch of Agnesi: an historical note on the
            Cauchy distribution, *Biometrika* **61** (1974), 375–380.
[Sto51]   M. H. Stone, On unbounded operators in Hilbert space, *J. Indian
            Math. Soc. (N.S.)* **15** (1951), 155–192 (1952).

[Sto90]   M. H. Stone, *Linear transformations in Hilbert space*, American Mathematical Society Colloquium Publications, vol. 15, American Mathematical Society, Providence, RI, 1990, Reprint of the 1932 original.

[Trè67]   F. Trèves, *Topological vector spaces, distributions and kernels*, Academic Press, New York-London, 1967.

[Trè06]   F. Trèves, *Basic linear partial differential equations*, Dover Publications, Inc., Mineola, NY, 2006, Reprint of the 1975 original.

[VFHN13]   F. Viens, J. Feng, Y. Hu and E. Nualart (eds.), *Malliavin calculus and stochastic analysis*, Springer Proceedings in Mathematics & Statistics, vol. 34, Springer, New York, 2013, A Festschrift in honor of David Nualart.

[vN31]   J. von Neumann, Die Eindeutigkeit der Schrödingerschen Operatoren, *Math. Ann.* **104** (1931), no. 1, 570–578.

[vN32a]   J. von Neumann, Über adjungierte Funktionaloperatoren, *Ann. of Math.* (2) **33** (1932), no. 2, 294–310.

[vN32b]   J. von Neumann, Zur Operatorenmethode in der klassischen Mechanik, *Ann. of Math.* (2) **33** (1932), no. 3, 587–642.

[vN35]   J. von Neumann, *Charakterisierung des spektrums eines integraloperators*, Actualités Scientifique Industrielles. Exposés Mathématiques, Hermann, 1935.

[vNS41]   J. von Neumann and I. J. Schoenberg, Fourier integrals and metric geometry, *Trans. Amer. Math. Soc.* **50** (1941), 226–251.

[VW76]   G. Velo and A. S. Wightman (eds.), *Renormalization theory*, D. Reidel Publishing Co., Dordrecht-Boston, Mass., 1976, NATO Advanced Study Institutes Series C: Mathematical and Physical Sciences, Vol. 23.

[Wig76]   A. S. Wightman, *Hilbert's sixth problem: mathematical treatment of the axioms of physics*, Mathematical developments arising from Hilbert problems *(Proc. Sympos. Pure Math., Northern Illinois Univ., De Kalb, Ill., 1974)*, 1976, pp. 147–240.

[Wig85]   A. S. Wightman, *Une perspective sur la théorie quantique des champs*, no. 131, 1985, Colloquium in honor of Laurent Schwartz, Vol. 1 (Palaiseau, 1983), pp. 175–185.

[Wig95]   A. S. Wightman, *Some exactly soluble models*, Infinite-dimensional geometry, noncommutative geometry, operator algebras, fundamental interactions (Saint-François, 1993), *World Sci. Publ.*, River Edge, NJ, 1995, pp. 338–362.

[YLTL18]   Y. Ying, Y. Lian, S. Tang and W. K. Liu, Enriched reproducing kernel particle method for fractional advection-diffusion equation, *Acta Mech. Sin.* **34** (2018), no. 3, 515–527.

[Yos95]   K. Yosida, *Functional analysis*, Classics in Mathematics, Springer-Verlag, Berlin, 1995, Reprint of the sixth (1980) edition.

[ZRK15]   M. Zheng, B. Rozovsky and G. E. Karniadakis, Adaptive Wick-Malliavin approximation to nonlinear SPDEs with discrete random variables, *SIAM J. Sci. Comput.* **37** (2015), no. 4, A1872–A1890.

Index

Printed in the United States
by Baker & Taylor Publisher Services